Inspection of Metals, Volume I:
VISUAL EXAMINATION

INSPECTION OF METALS by Robert Clark Anderson

Volume I: Visual Examination

In preparation:
Volume II: Destructive Testing
Volume III: Nondestructive Testing

Inspection of Metals, Volume I:

VISUAL EXAMINATION

ROBERT CLARK ANDERSON, P.E., FASM
Anderson & Associates, Inc.
Houston, Texas

AMERICAN SOCIETY FOR METALS
Metals Park, Ohio 44073

Copyright © 1983
by the
AMERICAN SOCIETY FOR METALS
All rights reserved

No part of this book may be reproduced, stored in a retrieval system, or transmitted, in any form or by any means, electronic, mechanical, photocopying, recording, or otherwise, without the prior written permission of the publisher.

First printing, May 1983
Second printing, November 1983
Third printing, February 1989

Nothing contained in this book is to be construed as a grant of any right of manufacture, sale, or use in connection with any method, process, apparatus, product, or composition, whether or not covered by letters patent or registered trademark, nor as a defense against liability for the infringement of letters patent or registered trademark.

Library of Congress Catalog Card No. 82-74087

ISBN: 0-87170-159-6

SAN 204-7586

Editorial and production coordination by
Carnes Publication Services, Inc.

PRINTED IN THE UNITED STATES OF AMERICA

Preface

This book has been written with a real purpose in mind. For the past thirty years the author has worked exclusively in the field of metallurgical failure analysis. Much time has been spent on the witness stand throughout the courts of this land. The serious injuries, loss of life, and property damage caused by metal parts failing is appalling. Hardly a day goes by that such a failure is not reported in the news media. Solutions to a few of these problems require painstaking investigative work. The cause of the majority of such failures is quickly revealed by *simple visual examination*. The remainder of the investigation is confirmation and documentation. Most failures should never have happened.

Although many inspection methods are documented in great detail in the literature, a quick search will reveal that few detailed references have been devoted to visual examination. Perhaps this is because, on the surface, it seems so simple and obvious. Experience has shown this is not so. Visual examination is truly the most important procedure available for inspection and for the analysis of the cause of failures. If the information presented in this book can help prevent just one such catastrophic failure, the book will have more than served its purpose.

No book is original material. It is the result of information gained from many written works of others blended with many years of experience. Wherever possible, an effort has been made to recognize these works and give credit. If this has been omitted in any case, it was by error and not on purpose.

Few if any books are prepared by individuals even though they are given credit. Many hours go into typing, proofreading, editing, printing, advising, preparation of photos and drawings, and just coexisting with a preoccupied author. To list the names involved with assisting in this preparation would require many pages. In two cases it would be remiss not to give specific credit. All the employees of Anderson & Associates Inc., Consulting Metallurgical Engineers, contributed in many ways. My wife, Aileen, typed, edited, proofread, and offered encouragement, keeping me going when work slowed down.

Contents

1 INTRODUCTION   **1**
    Defining the Scope   *1*
    Purpose of Visual Examination   *1*
    How Visual Examination Is Accomplished   *2*
      Examination With the Naked Eye   *2*
      Magnified Examination   *3*
      Sketching   *3*
      Thought Transference   *3*
      Markings   *3*
      Abuse   *5*
      Heat Effects   *12*
      Corrosion Scaling   *12*
      Cracking   *15*
      Fracture Examination   *16*
      Measurements   *18*
    Results and Record Keeping   *20*

2 EQUIPMENT   **21**
    Magnifying Devices   *22*
      Magnifying Power   *22*
      Focal Length   *22*
      Lens Types   *23*
      Simple Magnifiers   *27*

Lighting *31*
  General Lighting *31*
  Specific Lighting Devices *33*
Measuring Devices *33*
  Linear Measuring Devices *34*
  Reticles *35*
  Micrometers *35*
  Optical Comparators *37*
  Miscellaneous Measuring Devices *37*
Record-Keeping Mechanisms *39*
  Tape Recorders *40*
  Photography *40*
  Photography Lighting *43*
  Motion Picture Photography *43*
  Video Recorders *43*
Miscellaneous Equipment *43*
  Stereoscopic Microscope *44*
  Mirrors *45*
  Borescopes *45*
  Fiber Optic Scopes *48*
  Surface Finish Comparators *48*

3 EXAMINATION OF RAW MATERIALS AND FINISHED PRODUCTS 49

Defining the Categories *49*
Wrought Materials *50*
  Ingots *50*
  Blooms, Slabs, and Billets *50*
  Merchant Bars *50*
  Plates *51*
  Structural Shapes *51*
  Hot Strip Mill Products *51*
  Wire *51*
  Tubular Products *51*
  Defects *52*
  Ingot Defects *52*
  Defects in Blooms, Slabs, and Billets *55*
  Inspection of Plates *57*
  Structural Shapes *60*
  Inspection of Merchant Bars *60*
  Defects in Wire Products *62*

Defects in Tubular Products   *62*
Forgings   *64*
   Defects in Forgings   *64*
Castings   *68*
   Casting Defects   *68*

4 FABRICATION   79
Welding   *79*
   Visual Examination Prior to Welding   *80*
   Electrical Resistance Welding   *93*
Brazing   *99*
Soldering   *99*

5 INSPECTION OF MANUFACTURED PRODUCTS   103
Justification of Inspection   *103*
Specifications   *104*
Inspection at the Supplier's Facilities   *105*
Incoming Materials Inspection   *105*
In-Process Inspection   *105*
Finished Materials Inspection   *106*
Field Inspection   *106*

6 FAILURE ANALYSIS   109
Background Information   *109*
Tools of the Trade   *110*
Preparations for Sample Removal   *112*
On-Site Inspection   *112*
Examination Procedure   *113*
   Fracture Examination on Site   *114*
   Measurements, Sketches, and Notes   *116*
   Sample Selection   *117*
   Visual Fracture Examination   *119*
   Electron Microscope Examination   *123*
   Corrosion   *126*
   Fatigue   *130*
   Other Causes of Brittle Fracture   *130*
   Impact Fracture   *138*

BIBLIOGRAPHY   139

INDEX   143

CHAPTER 1

Introduction

Defining the Scope

Probably the single most important method of inspection of materials is visual examination. Visual examination here is defined as examination using the naked eye, alone or in conjunction with various magnifying devices, without changing, altering, or destroying the materials involved.

Visual examination can be usefully applied in many areas. Three of these which will be described in some detail are:

1. The materials used in the manufacture of other products (raw materials, semifinished materials, and components).
2. Manufactured products (during processing and finished products).
3. Defective or failed products (may or may not have been in use).

Purpose of Visual Examination

Visual examination is one of the least written about, least understood, and least effectively used of all procedures for inspection. There is a difference between looking at an object and really seeing it with a trained eye. Perhaps an analogy could be made to the medical profession. A physician has many sophisticated devices to aid in diagnosis and treatment. In spite of this, simple visual examination resolves most problems. The specialized equipment is used for confirmation of what has already been suspected.

Visual examination, on the surface, seems like a simple thing and one may wonder why a book would be written about it, or even how enough material for such a book could be gathered. It is hoped that this will be self-explanatory as the reader progresses.

How Visual Examination Is Accomplished

To do a good job of visual examination requires some knowledge of what you are looking at. It is good to have as much knowledge as possible of the product being examined. You must not only discover defects, but also be able to evaluate them from the point of usefulness or rejection. Knowledge of the cause of defective materials helps in future prevention. You should know what the inspected part is to be used for, how it is used, and perhaps how it may be abused. You should also be familiar with the types of defects which normally might be encountered in such a part, e.g., scabs, seams, and laminations in steel mill products; and corrosion, erosion, and physical abuse on parts that have been in service.

Examination With the Naked Eye

Start by looking at the part with the naked eye. This is the most difficult step. You must train yourself not to just look, but to see. A golfer of many years' experience will note that some people consistently find more lost balls than others. Observation has shown that these people look specifically rather than generally scanning the area. They walk, stop, and look as opposed to continually moving while looking. It is hard for the eye to see specific small objects while moving. Many veterans remember being taught that searching for ships on the horizon or for aircraft involved a very specific procedure. Instead of slowly moving your head to scan an area continuously, you moved your head a few degrees and stopped. Repeating this stop-start sequence made an amazing difference in what you could see.

Magnified Examination

After examination of all surfaces with the naked eye, proceed to some magnification. There are many magnifying devices available and these will be discussed in Chapter 2. The procedure will generally be to progress from the use of a simple magnifer to the use of a stereoscopic microscope. If still more magnification is necessary, the scanning electron microscope can be utilized.

Try to account for all unusual surface markings and conditions. Think in terms of depth effect and sharpness of penetration (stress concentration). Note any discoloration and determine the cause (heat? corrosion?). It should perhaps be pointed out that this book will be covering the dual role of examination for product inspection and examination for failure analysis. In both cases, determination of defectiveness and evaluation of such defectiveness are the goals.

Sketching

There are two things that have been found to be very useful and helpful in visual examination. It is a good idea to draw a picture of the defective part being examined. Make it as detailed as possible. When you draw a detailed picture of something you see many more characteristics. As a case in point, look at a person's face, then look away and try to remember details. Look at the same face and sketch it (how well you draw is not at all important). You will find that in sketching, many things will be seen that you did not otherwise note.

Thought Transference

The other useful procedure is to try to put yourself inside the part. Then imagine what sort of reaction must have occurred to have caused such a defect or failure. This sometimes gives a different perspective. Try to look at the parts and let them tell you their story. Avoid trying to impose your opinion on them.

Markings

Observe any identification markings. Such markings may identify the manufacturer, date of manufacture, original size of material, material specification, and number of the original heat of steel traceable to analysis and physical properties. If you are unfamiliar with identification marking procedures, do not assume

4 / Inspection of Metals: Visual Examination

Fig. 1 Stenciled identification markings on stainless steel plate.

a part is unmarked. Find out by asking manufacturers how they identify such products. Figures 1 and 2 show several such markings. Figure 1 shows stenciled markings on a stainless steel plate:

1/4 T304	(1/4 in. thick Type 304 stainless steel)
HT-825465-W12967A	(heat and slab number)
ASME SA240	(American Society of Mechanical Engineers Specification No. SA 240)
C.-.057	(carbon content)

Figure 2 shows:

WKM Co.	(manufacturer)
Houston Texas USA	(place of manufacture)
CWP 800 WOG	(continuous working pressure 800 lbs with water, oil, or gas service)
Steel	(material type)
24403	(pattern number)

These are obvious markings. But there are also many secret markings used by manufacturers, which they often do not readily reveal. One example is the wire rope industry. Some hemp center wire ropes have a single fiber wrapped with the group that reveals

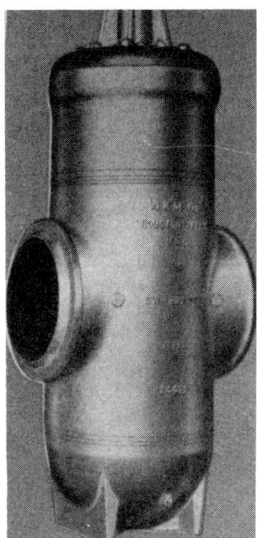

Fig. 2 Identification markings on a steel valve.

the name of the manufacturer and also the type of wire rope. It is difficult to distinguish this fiber and requires some care to separate and flatten it. Sometimes soaking in hot water helps (Fig. 3). Nearly all wire ropes have an identifying strand. It may be colored plastic, or the diameter of a single wire, or part of the construction configuration. Bolt heads are another good example of markings which give information. In addition to the radial markings on the heads which tell the strength level, the manufacturer's markings are often present (Fig. 4 and 5).

Abuse

Look for evidence of abuse. In the case of failure, try to decide if the abuse occurred before or after the failure. Failures often involve severe trauma. Parts flying about after the fact can suffer severe abuse that may be misinterpreted as being related to the cause of the failure (Fig. 6). If a part is distorted, try to ascertain the type, direction, and intensity of the load necessary to produce this distortion.

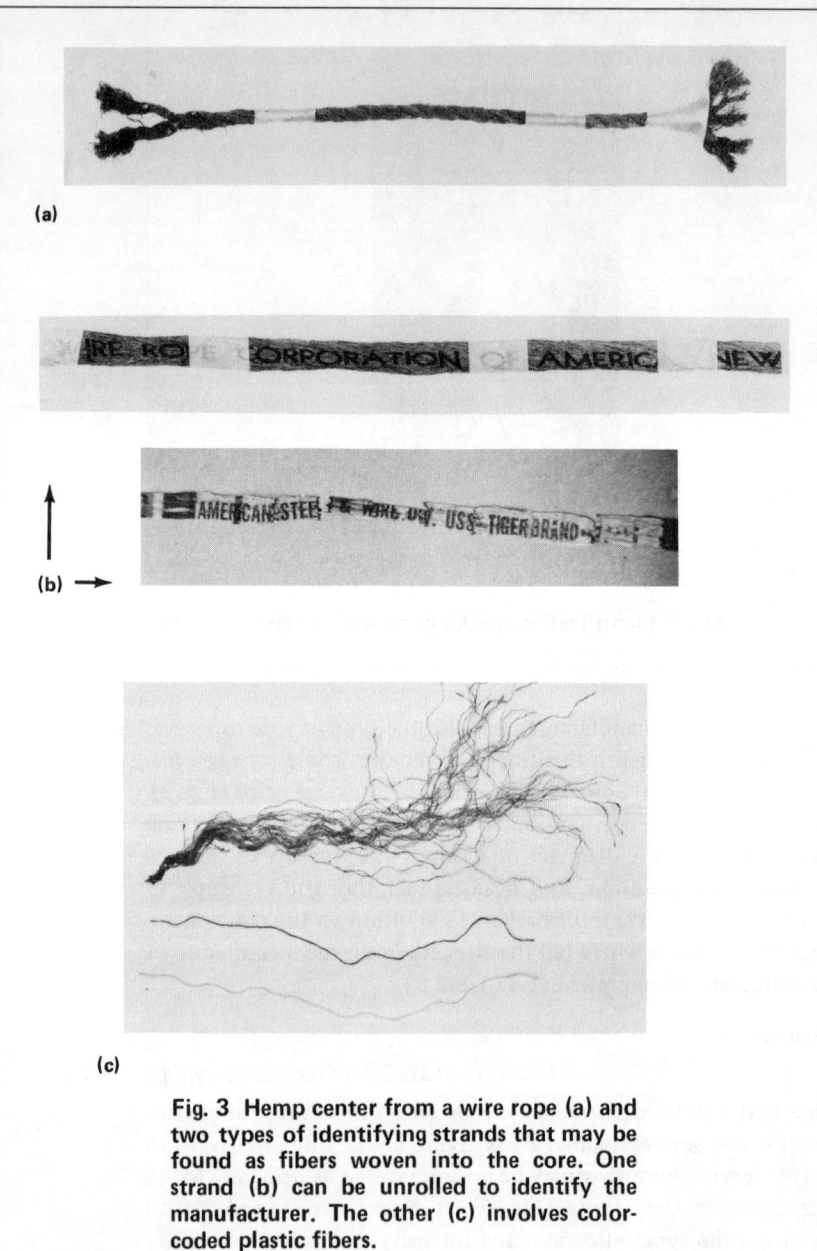

Fig. 3 Hemp center from a wire rope (a) and two types of identifying strands that may be found as fibers woven into the core. One strand (b) can be unrolled to identify the manufacturer. The other (c) involves color-coded plastic fibers.

Grade marking	Specification	Tensile Strength (min.), psi	Rockwell Hardness	
			Min.	Max.
 (plain hex)	SAE-Grade 1	60,000	B 70	B 100
	SAE-Grade 2	74,000 60,000	B 80 B 70	B 100 B 100
	ASTM A-307	60,000	B 69	B 100
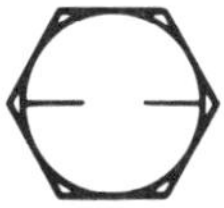 (two radial lines)	SAE 3	100,000	—	—
 (three radial lines)	SAE 5 ASTM A-449	120,000 105,000	C 25 C 19	C 34 C 30
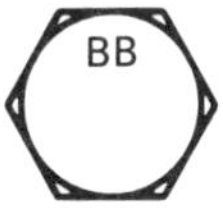 BB	ASTM 354BB	105,000	C 18	C 30
 A-325	ASTM A-325	120,000	C 23	C 34

Fig. 4 Chart showing bolt head markings. (Continued on next pages.)

8 / Inspection of Metals: Visual Examination

Grade Marking	Specification	Tensile Strength (min.), psi.	Rockwell Hardness Min.	Rockwell Hardness Max.
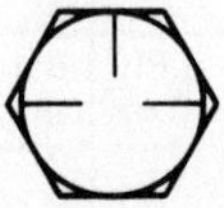	SAE-Grade 5.1	120,000	C 23	C 40
	SAE-Grade 5.2	120,000	C 26	C 36
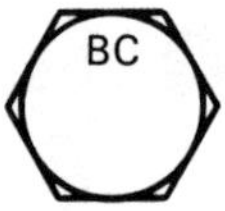 BC	ASTM A-354-BC	125,000	C 26	C 36
	SAE-Grade 6	133,000	C 28	C 36
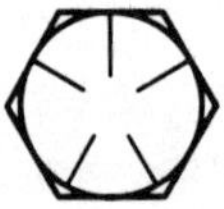	SAE 7	133,000	C 28	C 34

Fig. 4 Chart showing bolt head markings (continued).

Grade Marking	Specification	Tensile Strength (min.), psi	Rockwell Hardness Min.	Rockwell Hardness Max.
	ASTM A-354-BD SAE-Grade 8	150,000	C 33	C 38
	SAE 8.2	150,000	C 35	C 42
 A-490	ASTM A-490	150,000	C 32	C 38
	Exceeds SAE-Grade 8	185,000	C 36	C 40
 A-490	Specification underlined Atmospheric corrosion resistant	Varies	—	—

Fig. 4 (continued).

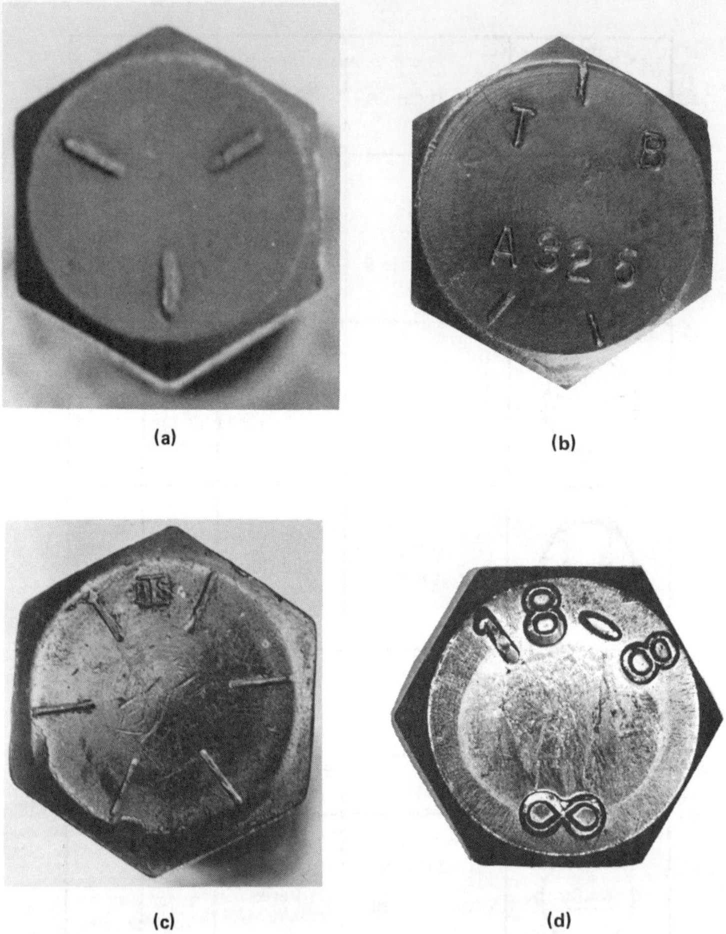

Fig. 5 Typical bolt heads showing type of material and manufacturer's symbols: (a) ASTM A449 or SAE Grade 5; (b) ASTM A325; (c) ASTM A354 Grade BD or SAE Grade 8; (d) Type 18-8 stainless steel.

Introduction / 11

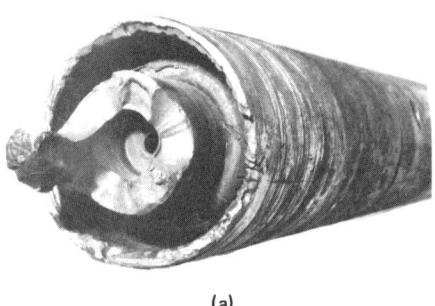

(a)

(b)

Fig. 6 Abuse and damage caused by parts breaking, then striking other objects: (a) a downhole oil field tool; (b) a turbine blade.

Heat Effects

Defects often relate to heating problems. These heating problems generally leave telltale indications. The indications are the result of the complex oxide systems of iron and other alloying elements. At lower temperatures, steel parts with a relatively bright surface may display characteristic temper colors. These colors vary somewhat with the composition and heat treatment. A variety of colors are produced within the temperature range of 380 F to 700 F. Charts such as the one shown in Fig. 7 (see insert between pages 54 and 55) can be consulted for approximate temperatures reached.

As the temperature rises, a heavier scale of a different type forms. The formation temperature varies with the type of alloy. Figure 8 shows the temperatures at which scaling becomes appreciable for various alloys. From this it can be seen that carbon steel, for instance, begins to scale appreciably starting at about 1050 F. It should be noted that a scale layer may be quite thick (Fig. 9). This does not necessarily indicate a great metal loss. Generally, the ratio of lost metal to oxygen in a scale layer is 8:1. This means a 1/8 in. thick scale layer represents only about 0.015 in. of lost metal. Since scaling can occur at such a low temperature, the microstructure may not have been altered appreciably (assuming the lower critical temperature to be 1333 F).

The color of such heat scale should also be observed. The brown- to red-colored scale layers on steel indicate the more completely oxidized Fe_2O_3. The black, mill-scale type of oxides are composed of the incompletely oxidized form of iron Fe_3O_4. This indicates an oxygen deficiency at the time of heating and may therefore mean much higher temperatures of formation. There are other temperature indications associated with observation of materials while being exposed to heat. Such materials just begin to glow dull red (in a dark area) at 1150 F to 1200 F. The colors change as the tempeature increases. For more accurate temperature measurement during heating, an optical pyrometer can be used.

Corrosion Scaling

Scaling of materials is not necessarily associated with heat. Corrosion also may create scale-type deposits. In some cases these may be indistinguishable from heat scale. Often, valuable data can

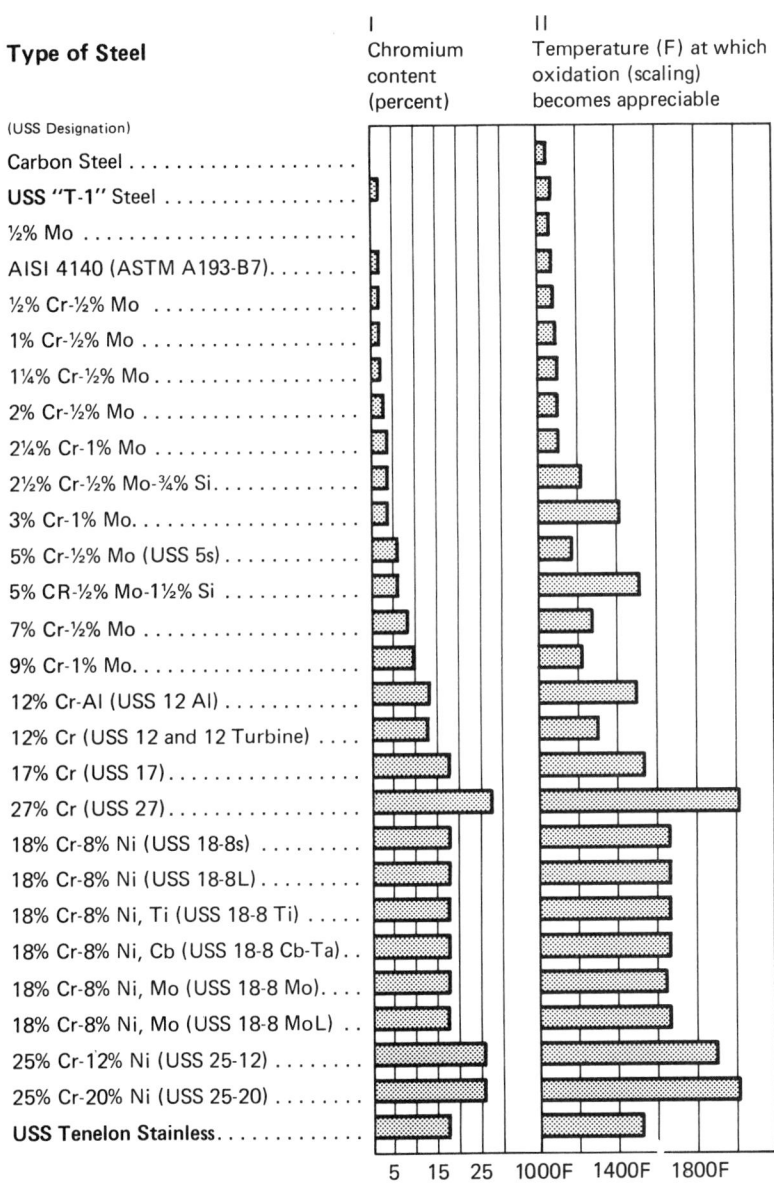

Fig. 8 Chart showing temperatures at which scaling becomes appreciable. (Courtesy U.S. Steel Corp.)

Fig. 9 Heavy scale layer formed at elevated temperatures.

be gained from scale analysis. It may be a good idea to remove scrapings and label for future tests. Try not to mix grease, paint, coating material, and mill scale with the corrosion product. If parts have internal and external surfaces, do not mix outside scrapings with inside scrapings. If corrosion is suspected, determine if it is localized or general. Is it uniform or selective (pitting)? Is it in an area of contact with other materials? Does it have special characteristics indicative of high velocity flow (e.g., erosion, cavitation)? Does it leave an unusual appearing corrosion product?

Cracking

If cracking is noted during visual examination, it is important to characterize the cracking. Is the crack straight or does it follow an irregular path? Is there one crack or a series of cracks? Are the cracks open or tightly closed? Are the cracks associated with markings of any kind? Are the cracks located in areas of natural stress concentration? Are the cracks associated with welding, for either fabrication or repair? Are the cracks associated in any way with evidence of corrosion, e.g., corrosion product, corrosion pitting?

If cracking is observed during the inspection of raw materials or in materials in process of manufacture or finished products, it should be explored for the purpose of determining acceptance or rejection. This is usually done by grinding to determine depth and extent. In such grinding, care must be exercised so that heat and stress do not cause extension of cracking. Grinding, if not done with care, can also close tight cracks by causing the adjacent metal to flow over the crack. It is often necessary to use methods such as magnetic particle testing or dye penetrant inspection to be assured that cracking is completely removed.

Once the crack has been completely removed, judgment must be exercised as to whether the part (1) can be used as is, (2) can be repaired, or (3) must be rejected and scrapped. Some specifications prescribe how much cracking is acceptable. The American Petroleum Institute, in many of its specifications, allows surface defects if the depth is no greater than $12\frac{1}{2}$ percent of the wall thickness. Where no such specification exists, the ultimate usage must be considered.

Of primary consideration is dynamic loading versus static loading. Other helpful considerations in making an acceptance or rejection are drawing tolerances for the part and design information such as safety factors.

Once a decision is made, it is necessary to determine if welding repair is necessary or if the part can be used as is, with the defects ground out. Welds may cause a new set of problems. If the weld can be avoided, feather the ground-out area to a generous radius to avoid stress concentrations. The smoother this surface is, the less likely it is to crack again.

Straight, sharp, open, single cracks (Fig. 10a) are usually associated with very high stresses and/or material of lowered ductility. They may be associated with suddenly applied (impact) loads. Open cracks may also be associated with locked-up internal stresses of a large magnitude. As an example, parts with high locked-up stresses may spring open when cut with a saw. Conversely, when locked-up stresses act in the opposite direction the cracks will be tight, due to compressive stresses (Fig. 10b), and the part might clamp shut on the saw blade when cutting. Cracks which are jagged in configuration (Fig. 10c) may be indicative of ductile tearing or may represent a separation at the grain boundaries (intergranular). Irregular-shaped cracking is usually indicative of slower crack propagation. Multiple cracks (Fig. 10d) are often associated with corrosion, stress corrosion, corrosion fatigue, thermal fatigue, or localized trauma.

It is important to note the location and orientation of cracks. Cracks are related to high stresses of either internal or external origin. Cracks will occur at the weakest point, usually a point of stress concentration.

Stress concentration is the nonuniform distribution of stresses in a loaded part. Stress concentrations take various forms (Fig. 11). They may occur as changes in section size. They may be in the form of a radius which can be gentle and smooth (good) or sharp and rough (bad). Purposely made markings should be suspect. These can take the form of rough machine grooves, coarse grinding marks, stamped numbers and letters, indentations (Fig. 12), holes, keyways, and scratches.

Another source of crack initiation is welding (see Chapter 4). Particular attention should be focused on welds, where cracks are in the general area, whether these be fabrication or repair welds. Look at the size (small welds are undesirable); look at the appearance (good welds look well made). Look for arc strikes and splatter (bad). Look for undercutting, which produces a stress concentration groove adjacent to the weld.

Fracture Examination

Whenever cracking results in a fracture, and if further failure analysis is anticipated, care should be exercised in protecting and

Introduction / 17

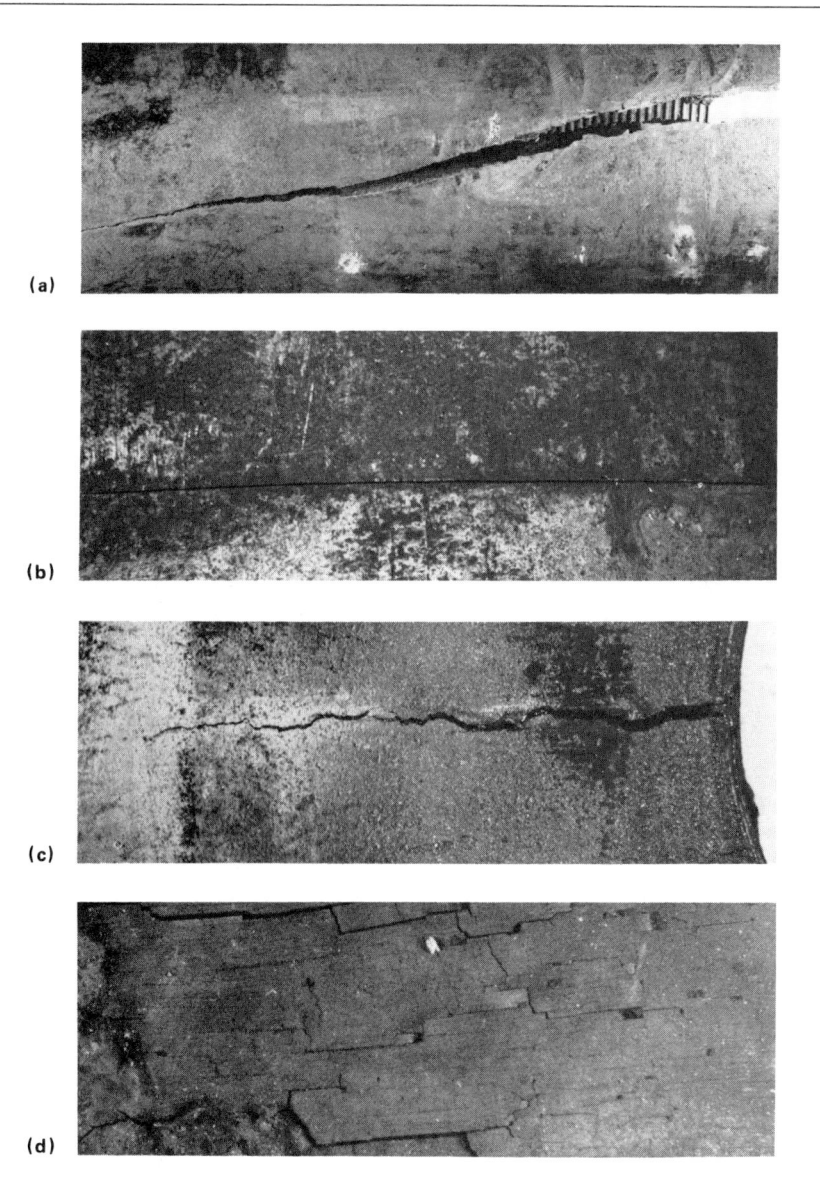

Fig. 10 Four types of cracks: (a) sharp and open; (b) sharp and tight; (c) jagged; (d) multiple.

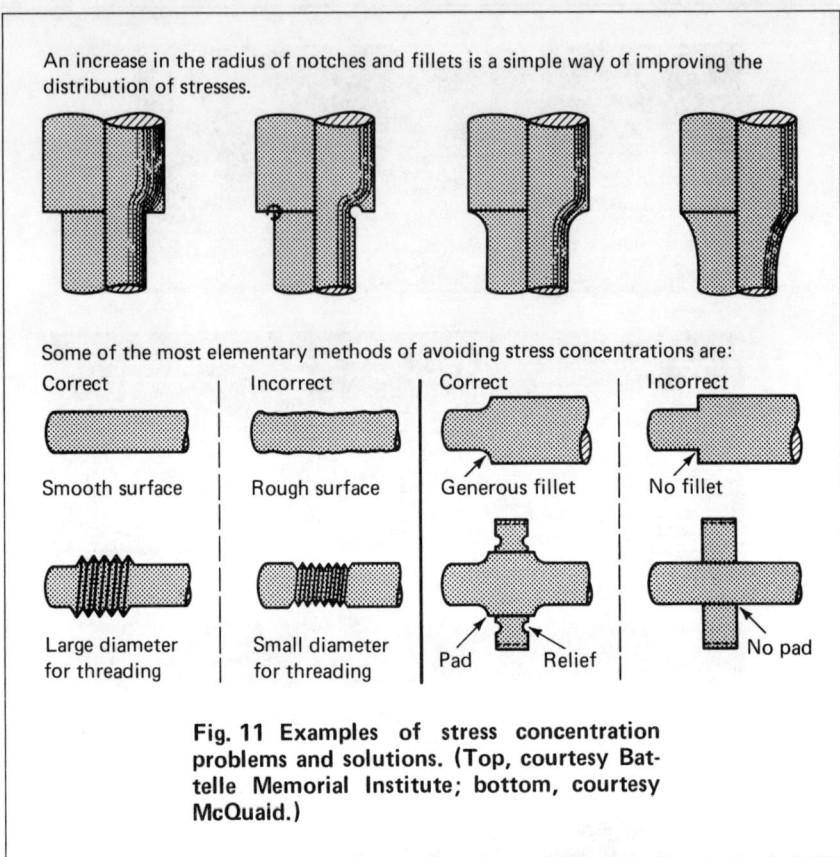

Fig. 11 Examples of stress concentration problems and solutions. (Top, courtesy Battelle Memorial Institute; bottom, courtesy McQuaid.)

identifying the fracture. Do not handle it any more than necessary. Do not fit the fracture surfaces together and do not clean these surfaces. This can alter the results of subsequent examination by distorting important fracture characteristics.

It is usually best to perform an overall examination prior to the fracture examination. Try to look at the parts in a general way and gradually become more specific. This can make the fracture examination more meaningful.

Measurements

Measurement is considered part of visual examination. Acceptance and rejection of materials may depend entirely on dimen-

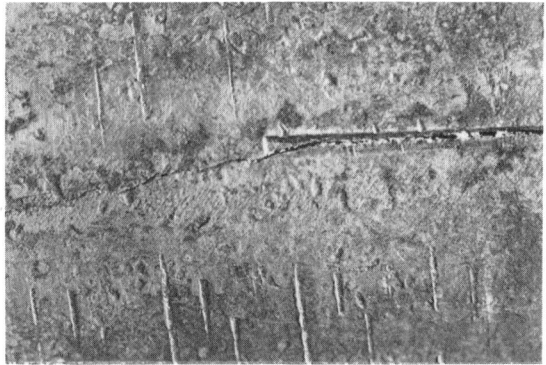

Fig. 12 Two examples of cracking caused by indentations that produced stress concentrations.

sions. Failure may also relate to dimensions. Most manufactured parts are made to some tolerance of measurement. Some of these may be to plus or minus several inches or even larger, while others may be as little as one ten-thousandth of an inch. Most are somewhere in between. If drawings and specifications are available, critical dimensions and tolerances can be readily determined.

There are numerous tools for measuring; these will be discussed in detail in Chapter 2. Measuring involves overall dimensions, inside and outside diameters, depths of holes, radii, thread sizes, and surface finishes. Depths of pits and lengths of cracks can also be measured if present. Most such measurements are simple, but some require very specialized equipment.

Results and Record Keeping

There must be a way to communicate the results of the visual examination, both to others and for future reference. We have already described the beneficial results of making sketches, which are one form of record. Written notes describing what has been observed are the most widely used method of record keeping. Printed forms with blanks to be filled in may aid the note-taking procedure. Dimensional measurements can be in the form of notes or as designations on sketches. Other common methods of record keeping are photography and verbal recording using a tape recorder. A good, clear, enlarged photograph is worth many words. Color photography should be considered. Not so commonly used are motion pictures and video tapes.

When recording results of visual examination, try to describe the part as if the reader had never seen it. Start by generally explaining what was examined, gradually becoming more specific. Describe the part and its condition thoroughly. This will assure good communication and lend credibility to your account. It will also serve as a good "refresher" if you are asked to explain your findings at a later date.

CHAPTER 2

Equipment

Visual examination is done best when you have knowledge of the existence of all the tools of the trade, knowledge of how to use the tools, and, of course, access to them. A "shade tree" mechanic, not having the specialty tools required for auto mechanics, soon discovers the job is at best very difficult and time-consuming, perhaps impossible.

Tools for visual examination can be grouped into five categories. *Magnifying devices* are essential. These may be very simple or quite elaborate and will be discussed in detail. *Lighting* for visual examination cannot be overemphasized. Lighting devices, like magnifiers, are available in a wide variety from general purpose to specialized. *Measuring devices* are for the most part familiar, but there are a number of highly specialized instruments. *Record-keeping devices* are not as extensive. The principal choices lie in the many types of cameras. And there is a variety of miscellaneous visual inspection devices.

It is recommended that anyone responsible for visual examination become familiar with all the available equipment. Different people favor different equipment. There are many manufacturers and numerous suppliers of such equipment. Most suppliers offer free catalogs which depict and describe the equipment. Some sources for such supplies and equipment are trade magazines, indexes of companies, and the Yellow Pages of the telephone directory. Key words to look under are inspection equipment, laboratory equipment, metallurgical suppliers, and testing equipment.

Before making a decision to purchase equipment, it is advisable to test it and discuss it with others who have used it, if possible. A supplier may be able to put you in touch with other users.

Magnifying Devices

Before detailing specific magnifying devices, a few words about optics are in order.

Magnifying Power

An object appears to increase in size as it is brought closer to the eye. In determining magnifying power, the true size of the object is what the image appears to the eye at 10 inches. The 10-inch value is used as a standard because this is the distance from the eye one usually holds a small object when examining it. Linear magnification is expressed in diameters. The letter X is normally used to designate the magnifying power of a lens, e.g., 10X.

If you could focus on an object at one inch, it would appear 10 times larger. Since we cannot effectively focus the eye at one inch, a lens may be used to do so. Magnification can thus be defined as the ratio of the apparent size of an object seen through a magnifier (known as the virtual image) to the size of the object as it appears to the unaided eye at 10 inches.

Focal Length

The focal length is the distance from the lens to the point at which parallel rays of light striking one side of a positive lens will be brought into focus on the opposite side. For lenses of short focal length such as we are discussing, light 30 to 40 feet away can be considered parallel. The focal length can be determined by holding a lens such that light coming through a window, for example, will allow the image of the window or other object to focus sharply on a sheet of paper held behind the lens. The distance from lens to paper will then be the focal length.

Once the focal length is known, the magnification of the lens can be determined, and vice versa. The shorter the focal length, the greater the magnifying power. The distance of the eye from the lens must be the same as the focal length. A lens with a one-inch focal length, for example, will have a magnifying power of 10 (10X). This is true if the lens is held one inch from an object and the eye is placed one inch from the lens.

In summary, the following formula determines magnifying power:

$$\textit{Magnifying Power (any positive lens)} = \frac{10}{\textit{Focal Length}} \textit{ (in inches)}$$

With a simple method of determining focal length, it becomes easy to determine magnification.

Lens Types

All lenses are either convex (bulged out), concave (sunken in), or flat. More often they are a combination of these. The most common type found in the laboratory is the double convex lens. Lenses with one side convex and the other flat (plano-convex) are used in projectors and microscopes. All other magnifiers are lenses used in combination.

There are three inherent faults in lenses—all of which are correctable. The degree of correction dictates the quality of the lens.

1. *Distortion*
   The image appears unnatural. The quality of the lens material and the grinding and polishing are both the causes of and the means for correcting this problem.
2. *Spherical Aberration*
   Light rays passing through the center of the lens and at the outer edges come to a focus at different points. (Naturally, the distortion is worse on large diameter lenses than small.) Spherical aberration can be corrected by slight modification of the curved surfaces.
3. *Chromatic Aberration*
   This is a prism effect: when broken down into colors, the light rays do not focus at the same place. This may occur both as a lateral and as a longitudinal effect. It is correctable by use of compound lenses of different types of glass.

Below about five magnifications one double convex lens is satisfactory. Two combined lenses will have a shorter focal length than either lens used alone. Higher magnification in simple magnifiers usually employs two or three lenses in combination. Twenty magnifications is about the maximum for these simple devices. A 20X magnifier will have a focal length and a field of view of about 1/4 inch.

When choosing a simple magnifier, consider the following corrected lenses for quality (Fig. 13):

1. *Coddington Magnifier* (Fig. 13a)
   This device uses a double convex lens with a groove ground in the middle. This diaphragm-like groove improves image quality by eliminating marginal rays of light.
2. *Double Plano-Convex Magnifier* (Fig. 13b)
   This two-lens magnifier gives partial chromatic correction and flatter field of view.
3. *Hastings Triplet Magnifier* (Fig. 13c)
   This is a multiglass lens corrected for both spherical and chromatic aberration. This is the best of all hand-held magnifiers.

The principal limiting factor for magnifying devices is depth of field. As magnification increases, the distance between the peaks and valleys (of an irregular surface) that are simultaneously in focus lessens. For example, at 100 magnifications the surface examined must be flat and polished. A variation of only 1/1000 of an inch can be out of sharp focus. The other factors, focal length and field of view, have already been discussed.

In summary, as the magnifying power of a lens system increases, (1) there are fewer peaks and valleys in focus at the same time, (2) the area observable is smaller, and (3) the distance from lens to subject becomes shorter (in addition, among other problems, lighting the object is difficult). Common magnifiers rated over 20X, while readily available, are therefore not very practical.

One other limiting factor in magnifying devices is light loss due to reflection. Lens surfaces can be coated with special antireflection coatings to reduce light loss, which may be particularly useful when the level of light is low. Figure 14 shows charts helpful in evaluating lenses.

Equipment / 25

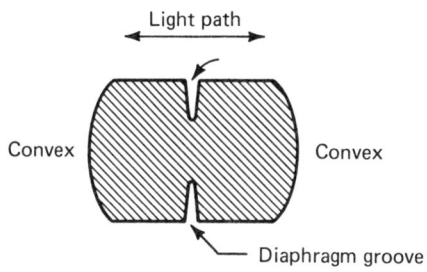

(a) Coddington magnifier

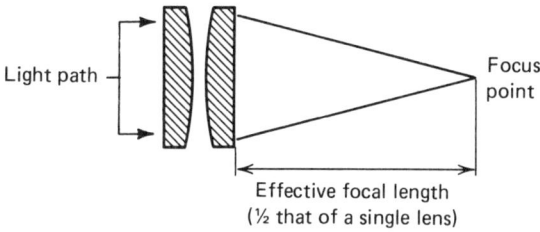

(b) Double plano-convex magnifier

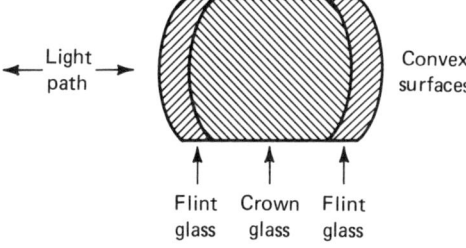

(c) Hastings triplet magnifier

Fig. 13 Diagram showing lens corrections available in simple magnifiers.

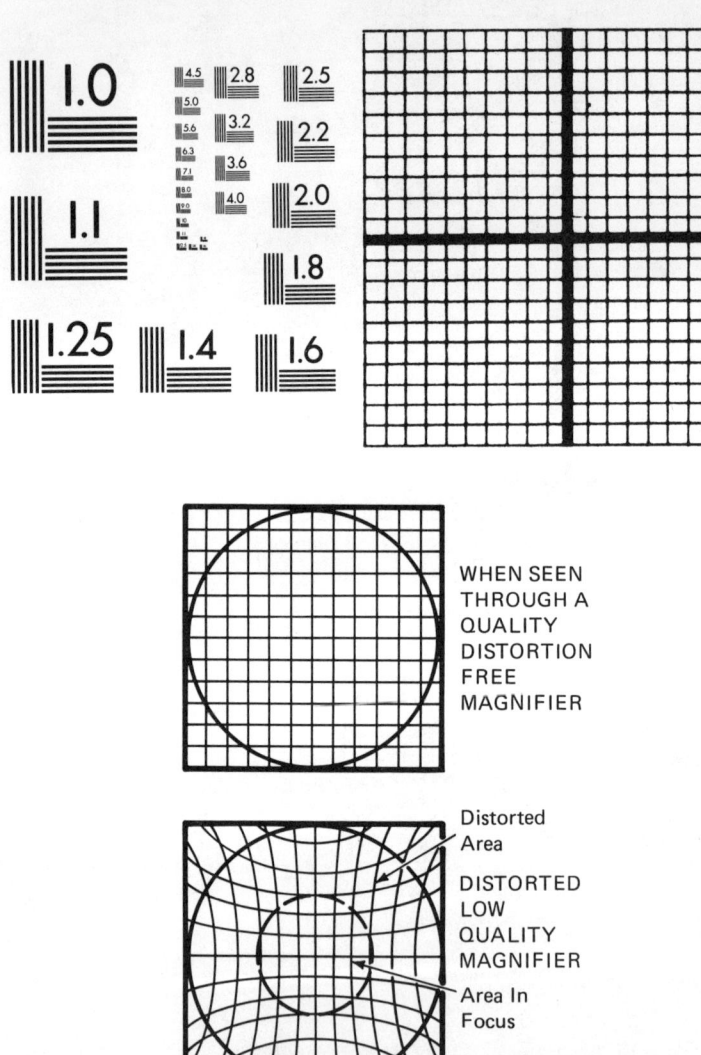

Fig. 14 Lens evaluation charts. (Courtesy Edmund Scientific Co.)

Simple Magnifiers

These come in many varieties, and new devices are regularly being developed. The following is an effort to group the various devices into categories:

1. Hand-held lenses, single and multiple.
2. Pocket microscopes.
3. Self-supporting magnifiers.
4. Magnifying devices which can be worn attached to the head or in some manner be used like eyeglasses or in conjunction with eyeglasses.
5. Magnifying devices with built-in light sources.

Hand-Held Lenses. These are available as a lens by itself, a lens with a frame and handle, or a lens that folds out or slides out of its own case. The fold-out type may include one to four lenses that can be used alone or in conjunction with one another. The size generally varies from 1/2 inch to six inches in diameter. They are available with either glass or plastic lenses.

The plastic (generally acrylic) lenses are shatterproof, but scratch easily. They are not capable of producing the lens corrections and quality of the glass lenses. The best of these hand-held lenses, as previously described, are the "Hastings Triplet," "Coddington," and the "Plano-Convex," in that order (see Fig. 13 and 15).

Pocket Microscopes. Another variety of the hand-held magnifier are pocket microscopes (Fig. 16). These are generally small-diameter tubes, about 1/2 inch in diameter and six inches in length, although they are also available in larger diameters. The smaller varieties are usually offered with magnification ranges of 25X to 60X. The subject end is cut at an angle or is somehow opened to allow maximum available light along with support. At these magnifications, the field of view and focal length are extremely limited, as is the available light. Auxiliary light is often a necessity. The larger-diameter units have lower magnifiying power.

Self-Supporting Magnifiers. Self-supporting magnifiers (Fig. 17) are much like the hand-held magnifiers, except they free the hands to manipulate the object being observed. They are generally low-power magnifying devices like the hand-held lenses. They are available as lenses with heavy bases and movable

28 / Inspection of Metals: Visual Examination

Fig. 15 Hand-held magnifiers.

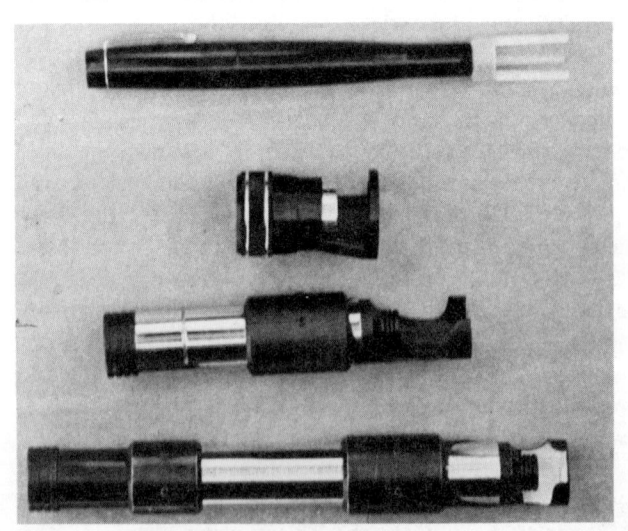

Fig. 16 Pocket microscopes.

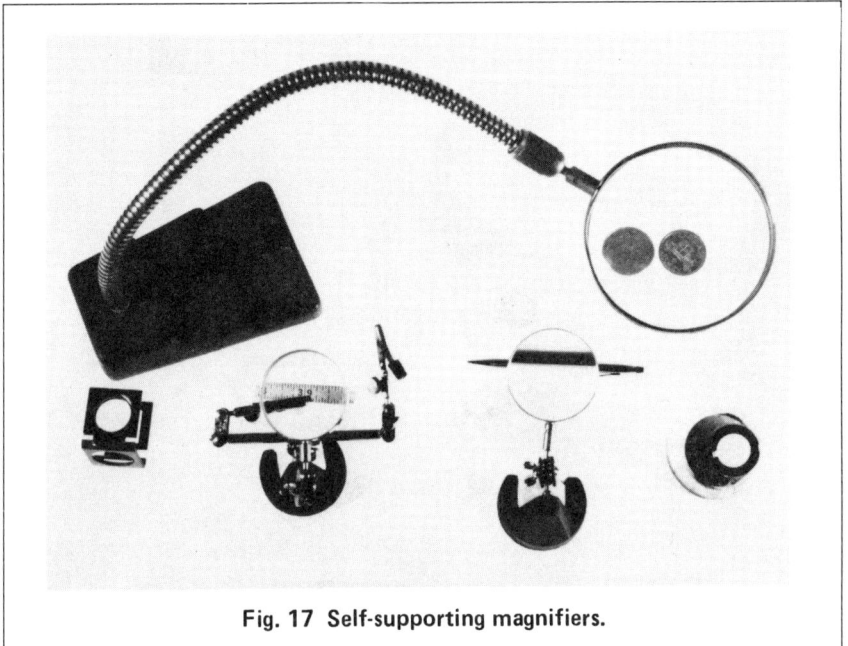

Fig. 17 Self-supporting magnifiers.

extension arms, lenses that sit directly on top of the object being viewed, and lenses that hang around the neck.

Magnifying Devices Which Are Eye Attachments. These magnifying devices are of two types. The visor type (Fig. 18a) has an adjustable band that fits over the head. This band supports a lens holder that tilts up and down for use when needed. The lens system may be two separate lenses or a continuous strip lens. It is also available with a loupe accessory for additional magnification. These visors may be worn with or without eyeglasses. Magnification offered is generally low (1½X to 3½X), but can be as high as 10X to 15X. They make excellent visual examination devices because they can be comfortably worn for long periods of time and can be quickly tilted in place for use when needed.

The second type is the loupe. Loupes used without glasses (Fig. 18c) either can be held in the eye by use of eye muscles, like a monocle, or are available with a spring clip which wraps around the head. Loupes are also available which attach to

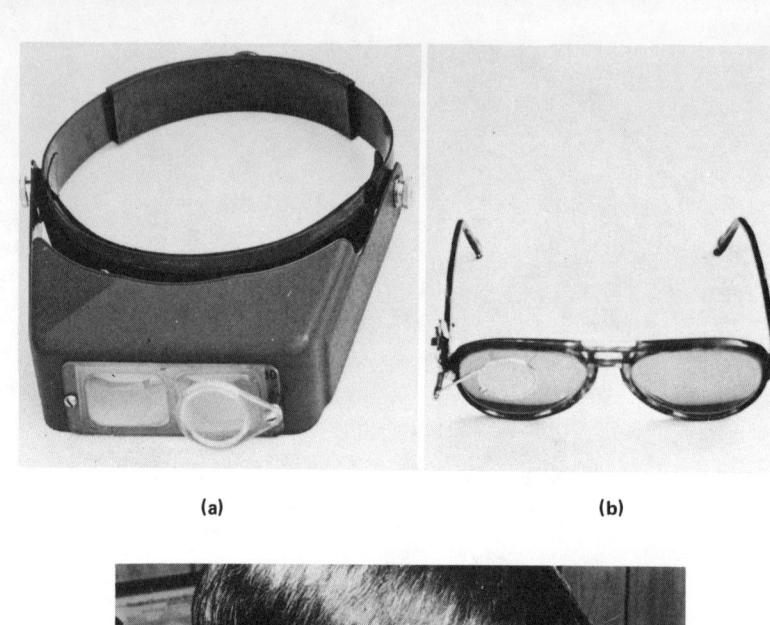
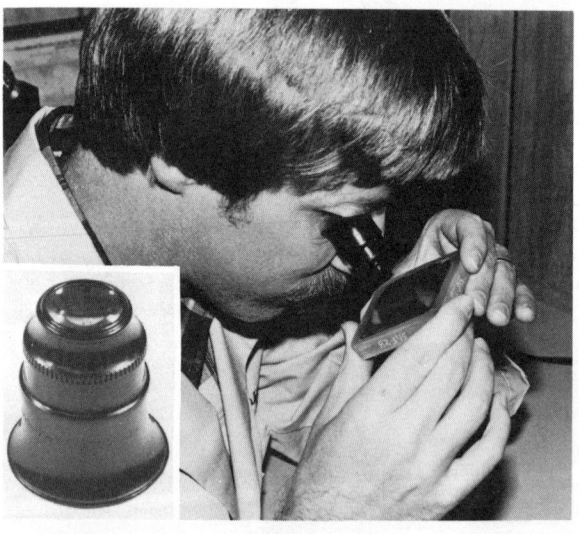

Fig. 18 Magnifying devices that attach to the head or eye: (a) visor; (b) eyeglass loupe; (c) loupe.

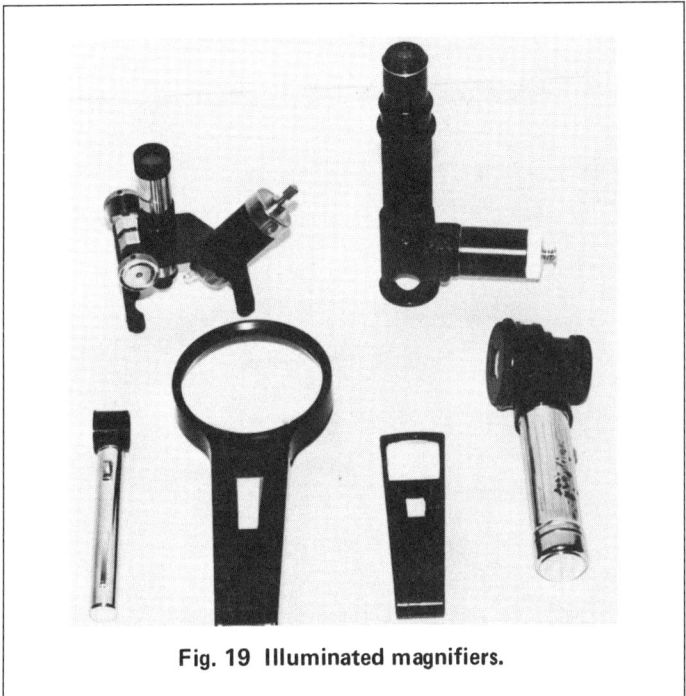

Fig. 19 Illuminated magnifiers.

eyeglasses as single or multiple lenses (Fig. 18b). These can be tilted in or out of use easily. The magnification range for such loupes is 2X to 18X.

Illuminated Magnifiers. Most of the magnifying devices described are also available with built-in light sources. To see details, good lighting is important. This is particularly true at the higher magnifications since the lens-to-subject distance is so short. Most light sources are either battery powered with flashlight batteries or equipped to plug into a standard wall outlet. The lights are usually incandescent, but are also available with fluorescent and ultraviolet light sources (Fig. 19).

Lighting

General Lighting

Very few indoor areas offer sufficient light to perform a proper visual examination. Sunlit areas are excellent for general examination, but may not be sufficient for examining internal

32 / Inspection of Metals: Visual Examination

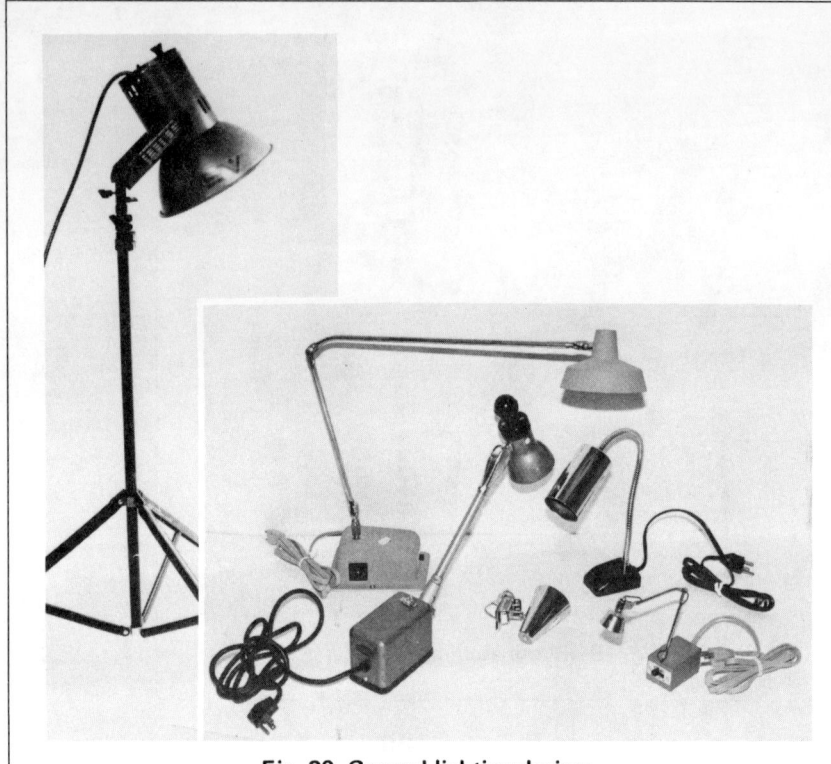

Fig. 20 General lighting devices.

areas such as bores and deep crevices. High-density fluorescent ceiling lighting offers good general-inspection lighting. For more specific overall lighting, there are three options (Fig. 20). One is a portable stand with an incandescent flood or spotlight bulb and reflector similar to those used by photographers. This gives a high-intensity source of light for a fairly large area. The stands are adjustable up and down, and the head swivels in all directions. This is a good light source for photographic recording. A word of caution on this type of light: bulb life is usually short (six hours), and considerable heat is generated.

When considering such equipment, it is wise to choose the sturdiest available. Two things to look for are heavy-duty swivel adjustments on the light head, and adequate cooling for the lamp base. These heavy-duty lights are available, but not as easy to find as the more common light-duty types. They are considerably more expensive, but easily worth the price.

The two other general lighting devices are swivel-arm incandescents and swivel-arm fluorescents. These come in a variety of shapes, sizes, intensities, and swivel-arm types. They provide less intensity and illuminate a smaller area than the flood or spot type described above. They are good for smaller areas and have longer lives. The fluorescent type has less intensity, but produces fewer shadows and is cooler operating. Many of the incandescent types have variable intensity controls. These lights can also be used in conjunction with magnifying devices.

Specific Lighting Devices

Specific lighting devices are of high intensity and permit the light to be concentrated on a small spot. Several types are shown in Fig. 21. The more common varieties are incandescent. They usually utilize an adjustable transformer and one or more diaphragms. They are on adjustable heads. These devices are most commonly sold as microscope lights. The problem with them is that they burn out and overheat easily, they do not have sufficient intensity, and they tend to produce an image of the light-bulb filament on the subject being illuminated.

There are several other devices for high-intensity, highly localized lighting. Two of these are like the microscope lights previously described; one uses the halogen very-high-intensity light source, the other uses the carbon-arc light source. The latter offers the brightest light of all the available sources, but requires adjustments and arc replacement. A third available unit is a fiber optic device. This allows highly specific, high-intensity light to be brought very close to an object, even in confined quarters. It is excellent for higher-magnification viewing and extreme close-up photography.

Measuring Devices

Measuring devices are considered part of visual examination because they are used to record the results of the examination. Visual examination, among other purposes, includes checking to see if parts meet dimensional specifications. These devices are

34 / Inspection of Metals: Visual Examination

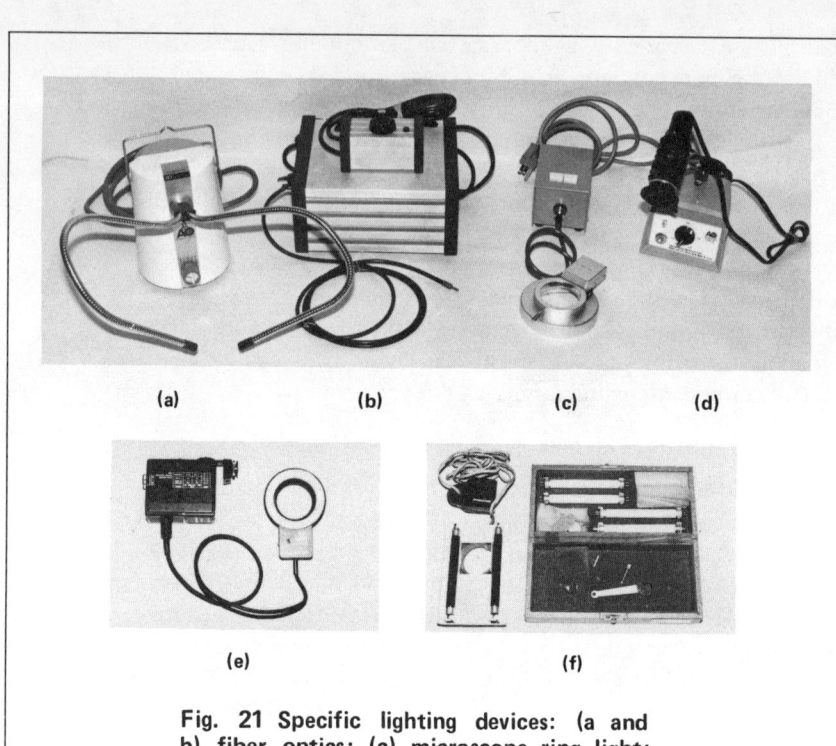

Fig. 21 Specific lighting devices: (a and b) fiber optics; (c) microscope ring light; (d) microscope light; (e) ring flash; (f) microscope illuminator using fluorescent and ultraviolet tubes.

so numerous, including many which are highly specialized, that a separate volume could probably be written about them. Because of this, only those most commonly used will be mentioned.

Linear Measuring Devices (Fig. 22)

The most common measuring unit is the ruled straightedge. These come in many forms, such as the 12-inch rule and 36-inch yardstick, both of which are gradually being replaced by metersticks. Until the transition to the metric system is universal, both English and metric devices must be utilized. In addition, tape measures, which are available from six inches in length to over 100 feet, are essential for visual examination.

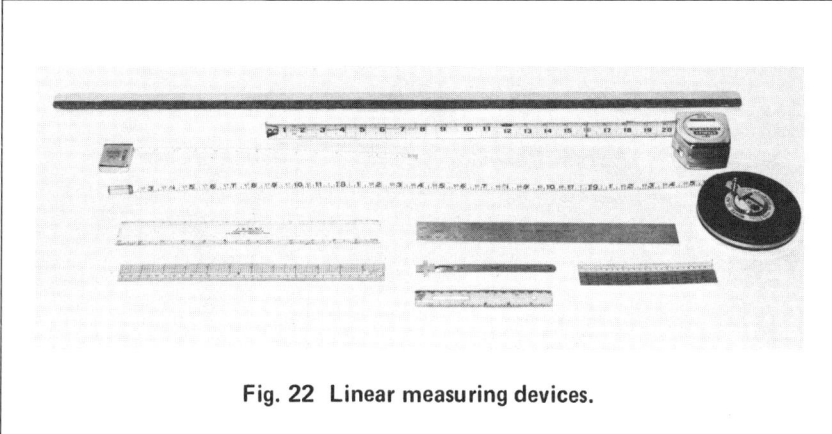

Fig. 22 Linear measuring devices.

Of the rules, the 6-inch and 12-inch steel rules are desirable. The six-inch scales, some of which can be clipped to a shirt pocket, are available with several scales on each rule. Scales can be both English and metric on the same rule and may be subdivided to as small as 1/100-inch divisions. These devices are also available with an adjustable 90° squaring edge to check for straightness.

Reticles

There are many magnifying devices available with built-in reticles (Fig. 23). Reticles made to measure nearly anything imaginable are available. If the desired reticle is not available, it can be custom-made.

Micrometers

Micrometers (Fig. 24) are extremely accurate mechanical devices. They are commonly used to measure to 1/1000 of an inch and can be used to 1/10 000 of an inch. Both inside and outside micrometers are available. Some incorporate both in the same unit. They are available with various measuring tips (these are normally hardened to prevent wear). Tips can be flat, rounded, pointed, or blade. Others may also be available or custom-made. Very little training is required to use these devices, but experience produces more consistently accurate results.

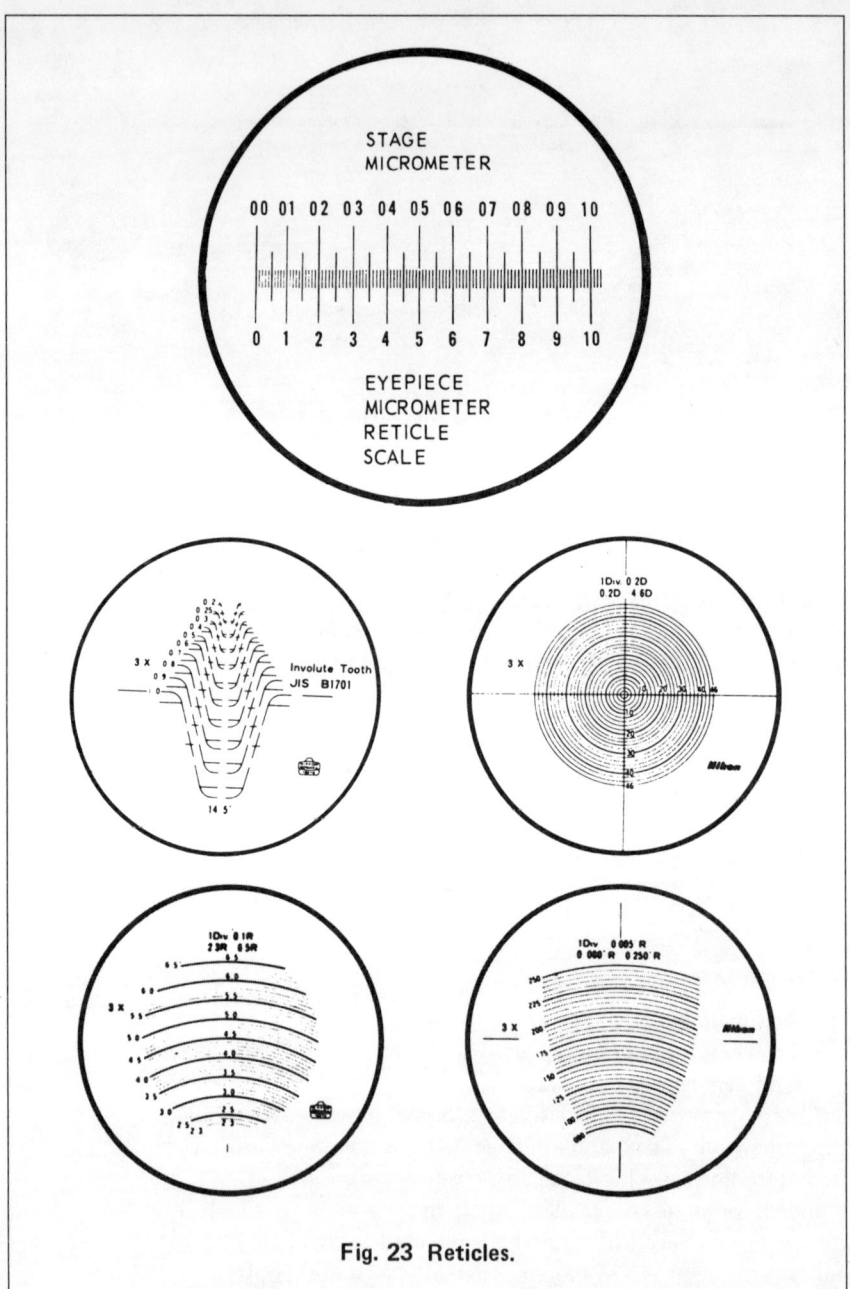

Fig. 23 Reticles.

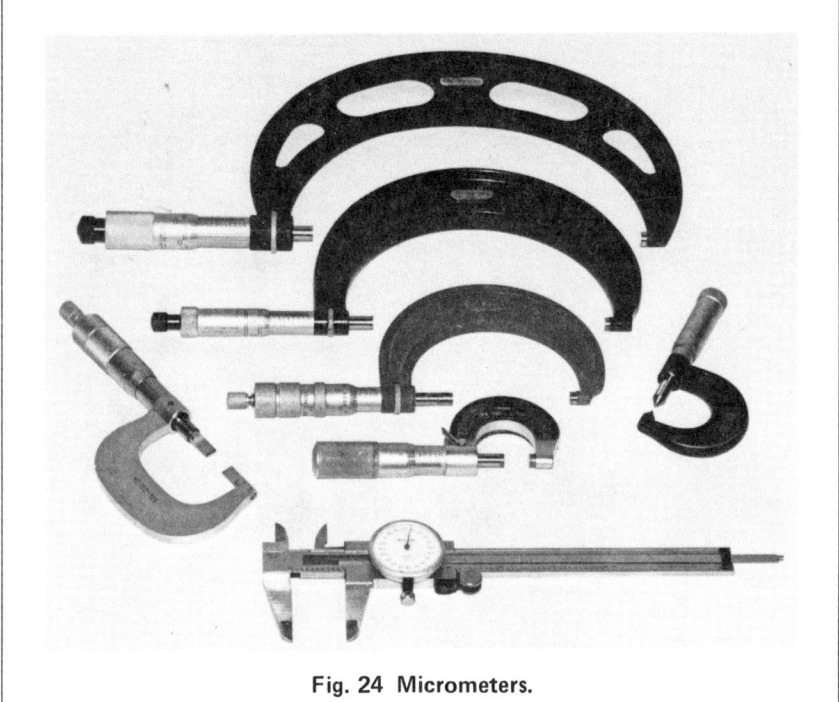

Fig. 24 Micrometers.

Optical Comparators

Optical comparators (Fig. 25) are excellent devices for both visual examination and measurement. A comparator produces a two-dimensional enlarged image of an object on a large ground-glass screen. It can be used with reflected light or background lighting (or a combination of both). Magnification is available from actual size to 50X. Comparison templates can be placed on the screen to check dimensional accuracy. Results can be readily photographed.

Miscellaneous Measuring Devices (Fig. 26)

The dial indicator consists of a plunger-actuated dial, usually calibrated in 1/1000 of an inch. It comes with a series of mechanical arms and clamping devices such that it can be attached to a fixed (rigid) object and reference measurements can be made.

38 / Inspection of Metals: Visual Examination

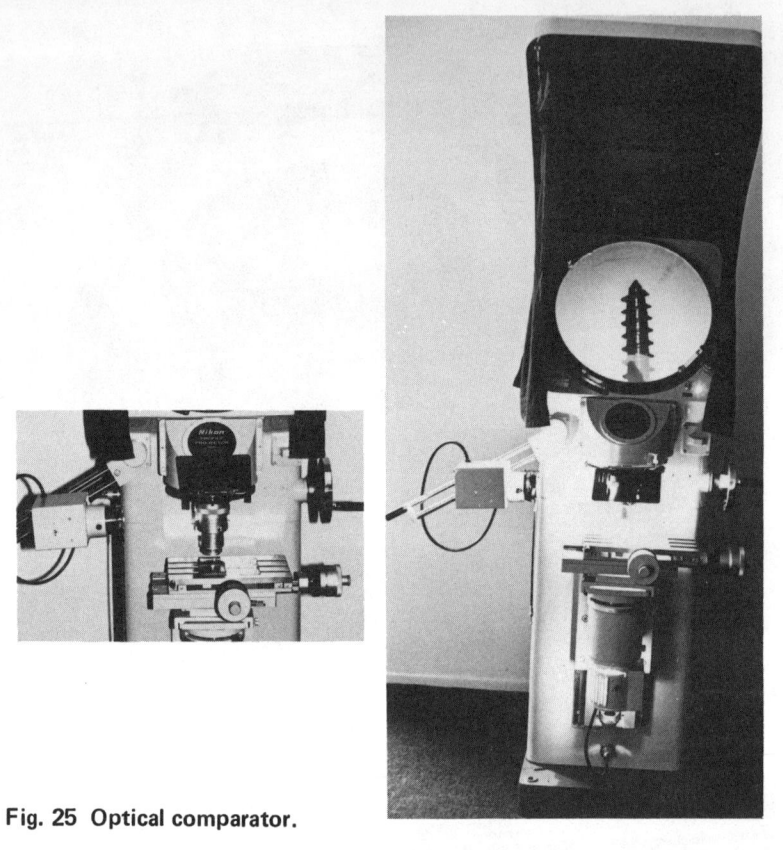

Fig. 25 Optical comparator.

A common usage is attachment to a lathe bed to check both horizontal and circumferential dimensional variations.

There are many specialty measuring gages. Some of these are inside micrometers, tubing wall measuring micrometers (one rounded anvil and one flat anvil), depth gages, thread-measuring gages, protractors and bevel protractors (to measure angles), levels (to measure variation from horizontal), inside and outside calipers, hole and plug gages (to measure diameter and uniformity of holes), radius gages, screw-pitch gages, thickness gages (a series of leaves of various known thicknesses to check clearance).

Equipment / 39

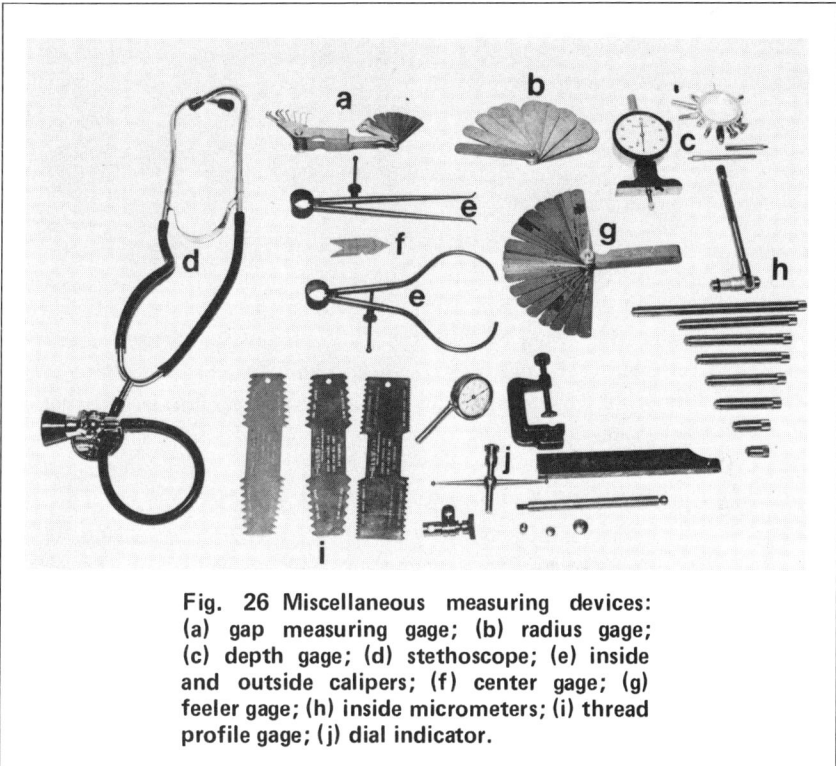

Fig. 26 Miscellaneous measuring devices: (a) gap measuring gage; (b) radius gage; (c) depth gage; (d) stethoscope; (e) inside and outside calipers; (f) center gage; (g) feeler gage; (h) inside micrometers; (i) thread profile gage; (j) dial indicator.

Many such devices are specially designed and built for a particular application. A wide assortment are available as stock items, with many brands to choose from. Quality should not be sacrificed for cost on measuring devices. They should be kept in specially designed cases, be kept clean, and be lubricated as required. When making measurements, devices should not be forced or overtightened. Many of these tools come with calibration blocks and should be checked regularly for best results.

Record-Keeping Mechanisms

Verification of the results of visual examination requires some means of record keeping. Record keeping in its simplest form is accomplished by making written notation of the results. Since the making of written records is a somewhat slow and inconvenient process and may indeed end up being illegible, other methods are worthy of consideration. Several devices for record keeping are shown in Fig. 27 and are described below.

Tape Recorders

One simple and widely used device is the pocket-type tape recorder. With it, rapid note taking is possible, and the results can be transcribed later. The most popular type is battery-powered, using cassette tapes. The tapes can be used repeatedly or can be identified, replaced, and preserved. Although tape recorders are simple and accurate, they cannot record sketches or visual depiction of results.

Photography

Photography is an excellent record-keeping method. It can be used very effectively in conjunction with written records or the tape-recorder. There have been many advances in this field in recent years, and volumes are available on the subject. The most commonly used devices for recording the results of visual examination are the 35mm single lens reflex camera, the Polaroid type of cameras, the larger-format view cameras, and macro cameras. Because each of these devices has advantages and disadvantages, it is difficult to recommend only one for a general-purpose device. For this reason, each will be discussed in detail.

35mm Cameras. The 35mm single lens reflex cameras may be the most widely used of the group. They are available with many accessories and lens types. Many have built-in light meters. They are very portable and produce excellent results. The operator looks directly through the lens in composing the picture, which prevents forgetting to remove the lens cap or having a thumb in the photo. At the time of exposure, a mirror device pivots out of the way, the exposure is made, and the viewing mirror returns to position. Lenses are interchangeable. Exposure times are variable and can be as little as 1/1000 of a second. Apertures vary, allowing light-entrance control and variation in the depth of focus. Lighting can be sunlight, artificial, flash, or strobe-type high-intensity lighting. These cameras usually take 20- or 36-exposure films. Film is available in many types and speeds, in

Equipment / 41

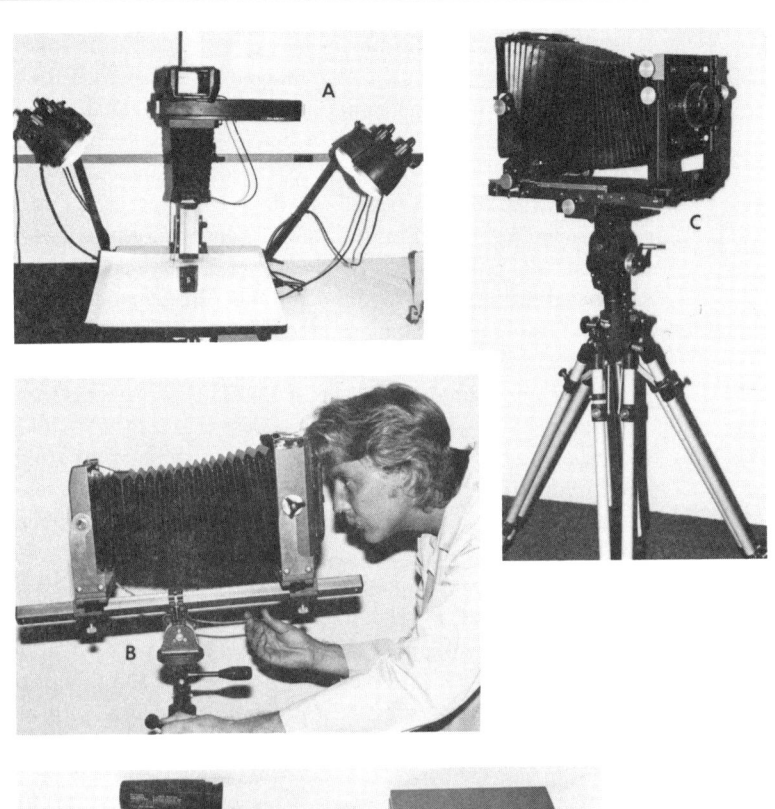

Fig. 27 Record-keeping devices: (A) macro camera; (B) 4 × 5 view camera; (C) 8 × 10 view camera; (D) video camera; (E) 35mm camera; (F) tape recorder; (G) motion picture camera.

either black-and-white or color. Film is also available for prints or slide projection. Cost per picture is low due to the small size of the film. The principal disadvantage of this camera is the small size of the film, which requires enlarging. If enlarged too much, loss of detail and graininess can occur. Another disadvantage is that 20 or 36 exposures must be made before developing, to avoid wasting film. After all of the exposures, delay may be experienced while the film is developed and each frame separately enlarged and printed. But the speed, versatility, portability, and variety of high-quality equipment and film generally outweigh any disadvantages, making 35mm cameras an excellent choice for a wide range of record-keeping chores.

Rapid Photography Cameras. Polaroid-type cameras have the advantage of producing rapid records. This ensures that the results are what was expected without the delay of shooting, developing, and printing a roll of film. There are various sizes of cameras, the most common being 4 X 5 inch. More recently, 8 X 10 inch color film has become available, adding a new dimension to rapid photography. Film is available in single sheets and film packs; in black and white; in both normal speed and high speed; with or without negatives. The film type with negatives produces a rapid print as well as a negative for additional copies. Instant film is also available in color. Camera backs for instant film are also available and can be used in conjunction with many conventional cameras.

View Cameras. View cameras are simply cameras with a larger format. Common sizes are 4 X 5 inch, 5 X 7 inch, and 8 X 10 inch. These cameras produce probably the best quality, most detailed photos. They are somewhat bulky and often require tripod support. Film is available in black and white and in color. Prints can be greatly enlarged with little loss of detail.

Macro Cameras. Macro cameras are another form of the view camera. They are particularly adapted to produce magnified photos. This is accomplished by use of special lenses in conjunction with focusing bellows. These cameras are best used with smaller parts. Lighting for this equipment is important. At higher magnifications, the lens-to-subject distance is short and lighting is even more difficult. A variety of films are available for use with this equipment, including black-and-white cut film, color film, and all of the rapid film types.

Photography Lighting

Lighting for all types of photography is very important. The choices are many, such as daylight, artificial light, floodlights, photoflash, and all of the highly specific lights described earlier under "Lighting." The type of light required is a function of the film type and, to a lesser degree, of the camera.

Motion Picture Photography

An advance in record-keeping photography is the motion picture or movie camera, which allows the taking of a sequence of thousands of photos. It records motion and when played back can slow or stop motion at any point. It can be used in conjunction with sound recording which becomes an integral part of the film. Sound can be the actual sound occurring during filming, or sound added later while viewing the film, as commentary. The most common film sizes are 8mm and 16mm, but both smaller and larger sizes are available and are used. Although this equipment can be very elaborate and expensive, adequate 8mm sound cameras and projectors are available at very moderate prices. Both black-and-white and color film are available.

Video Recorders

A record-keeping device that promises to be widely used in the future is the video tape recorder. The equipment has been used for many years by the television industry, but has only recently become available in more portable and inexpensive forms. The video tape recorder produces records similar to the movie camera. The principal difference is that the results can be viewed instantly without processing. They can be shown on any standard television or video screen, and the recording tapes can either be stored or erased and used again. The principal undesirable factor is that the image may not be as sharply detailed as other photographic recording.

Miscellaneous Equipment

There are many other tools of the trade for visual examination. In this book, we have tried to separate visual examination from nondestructive testing. There are several other notable items that are definitely classified as visual examination items.

44 / Inspection of Metals: Visual Examination

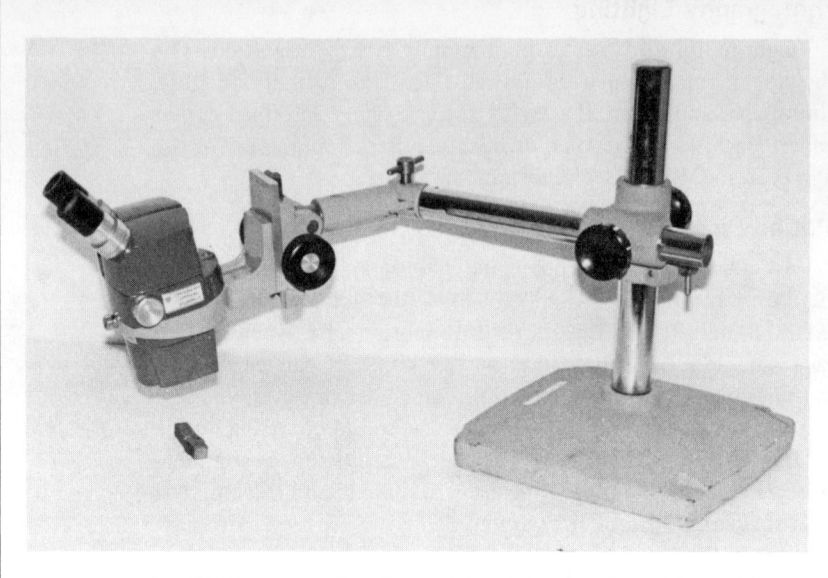

Fig. 28 Stereoscopic microscope with extension arm.

Stereoscopic Microscope

The stereoscopic microscope may well be the most important and widely used of all visual tools (Fig. 28). It allows three-dimensional viewing, clearly and sharply, to magnifications as high as 180X. There are several variations to consider with this equipment. These are lens combinations, the stand, and the zoom option. The lens combinations may be wholly or partly interchangeable. The ultimate use dictates the final choice, but for general-purpose work magnifications in the range of 5X to 50X are most popularly used.

The normal stand is similar to that for any upright microscope and is adequate if small parts are to be viewed. For most applications, however, the extension-arm type of stand is much more versatile. The long extension arm allows the "scope" to swing out over fairly large parts to examine a specific area.

The zoom option is highly recommended. It allows a continuously variable range of magnification without changing lenses, by simply rotating a dial. For example, magnification may be varied from 5X to 50X. General observations can be made at 5X, and if something of interest requires more detail, a higher magnification, up to 50X, can rapidly be obtained.

Camera attachments are available with stereo-microscopes, but unless stereo pair photography is used the results are disappointing compared to the visual observation.

Mirrors

Another essential tool for the visual examiner is a mirror. Mirrors are available in all sizes and shapes, with and without lights. They are available with long extensions, swivel heads, and remotely actuated heads. Several types are shown in Fig. 29. Mirrors are sometimes the only method for viewing inaccessible areas.

Borescopes

For inaccessible areas, a borescope (Fig. 30a) is helpful. Rigid-type borescopes are tubes of varying diameters with built-in lens systems. They generally have built-in lighting systems. They come in fixed lengths as well as in the sectional form. Lengths vary from 6 inches to about 40 feet. Diameters will usually vary from about 1/10 of an inch to one inch. The viewing heads

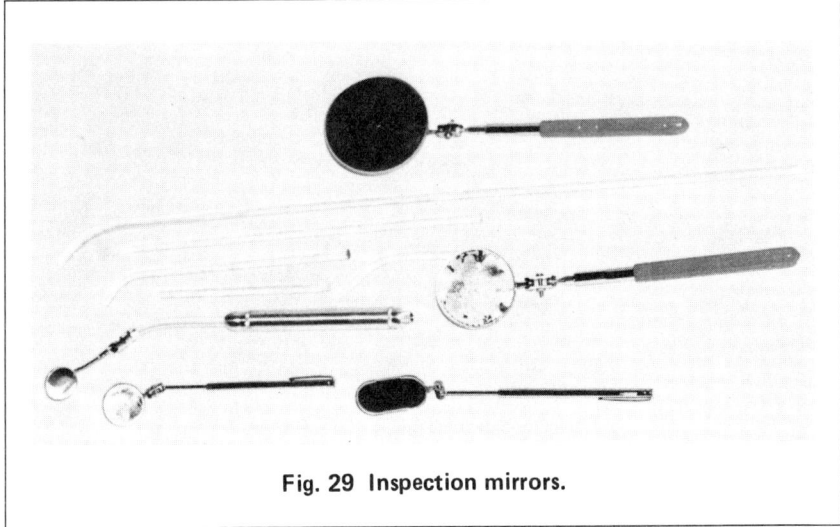

Fig. 29 Inspection mirrors.

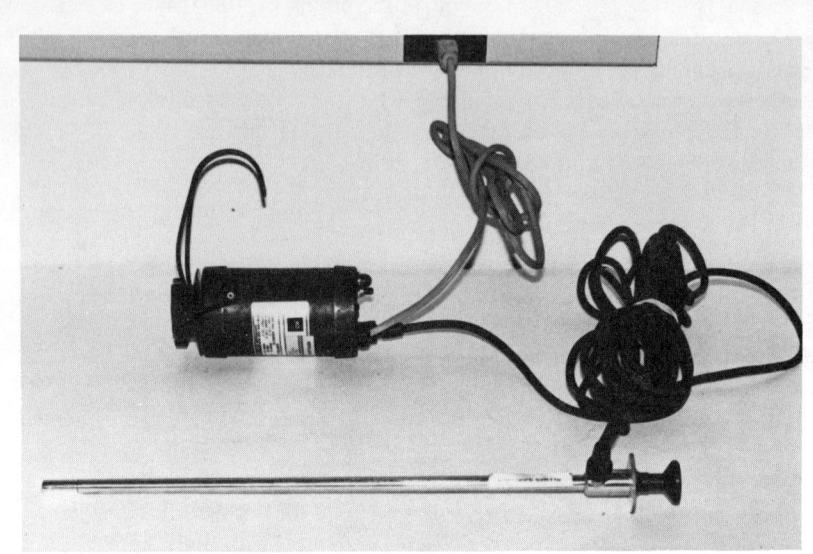

(a)

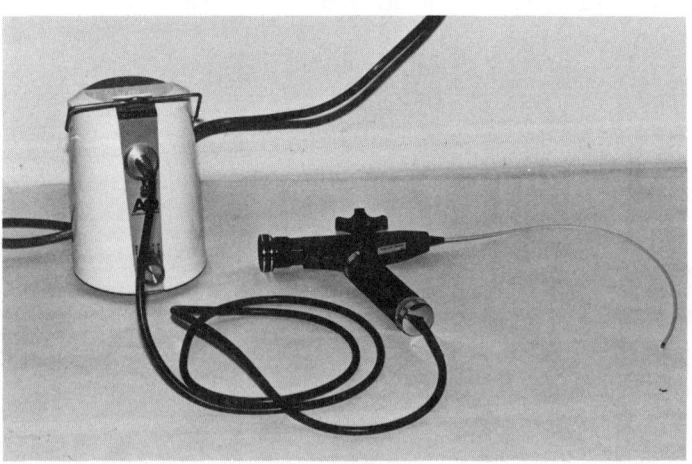

(b)

Fig. 30 (a) Borescope; (b) fiber optic scope.

Equipment / 47

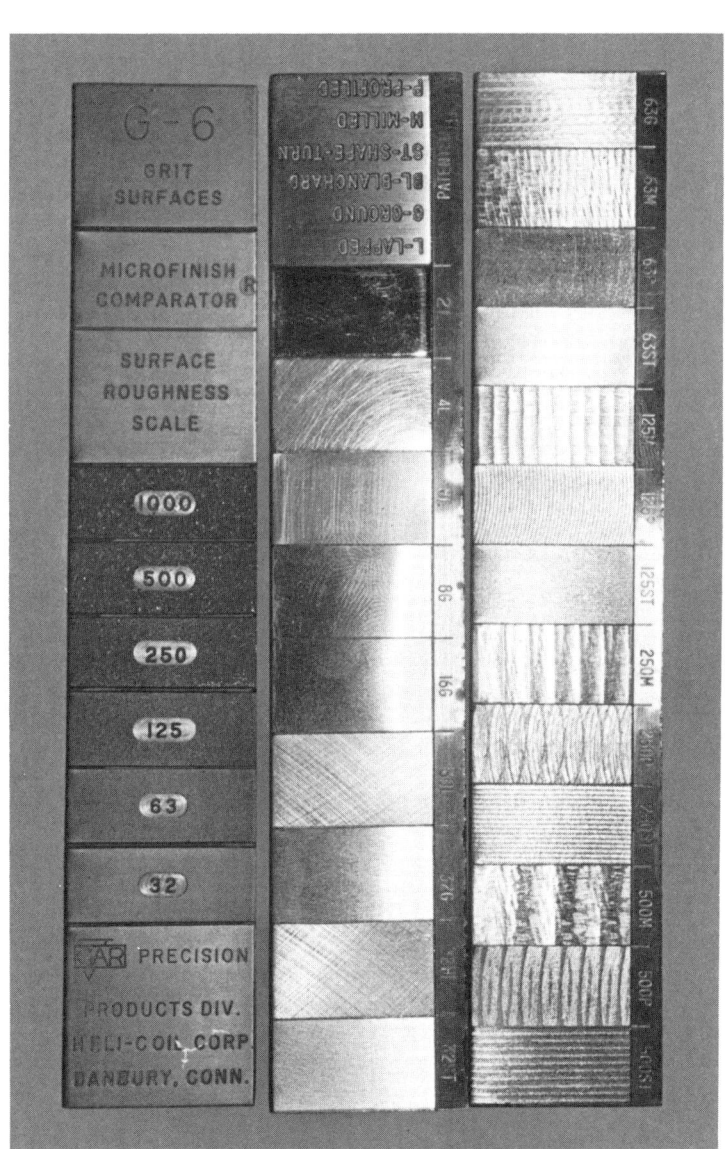

Fig. 31 Surface finish comparators.

offered are, for example, for straight-ahead viewing (0°), forward oblique viewing (30°), right-angle viewing (90°), retrograde viewing (110°), and panoramic viewing (180°).

Eyepieces are available in monocular or binocular viewing. They come in both fixed focus and adjustable eyepiece focusing. They are also available with adaptation to video viewing on television screens and for photography. Both incandescent and halogen light sources can be utilized. Either plug-in or battery power sources can be used. Magnification can be varied by design, but is principally related to the distance between the subject and the objective head. These instruments are excellent for internal examination of long tubes, boreholes, internal combustion engine cylinders, castings, etc.

Fiber Optic Scopes

Another variety of the borescope is the fiber optic scope (Fig. 30b). These are very similar to borescopes, but have the ability to deform. The examining tube in this case is made up of thousands of carefully aligned glass fibers with an objective lens at one end and a magnifying eyepiece at the other. Since it is flexible, a fiber optic scope can "snake" its way around corners and along tortuous paths to examine inaccessible areas rigid scopes could not reach. These scopes are available up to 15 feet long with a variety of accessories, including watertight viewing tubes.

Surface Finish Comparators

Many comparative test sheets are available to rate surface finishes (Fig. 31). The surface finish is often a requirement of visual examination. These comparator scales are available for rating machined surfaces, including turned, ground, lapped, milled, profiled, and EDM (electrical discharge machined). They are available for rating grit-blasted and sandblasted surfaces. Cast surfaces can also be so rated.

CHAPTER 3

Examination of Raw Materials and Semifinished Products

Defining the Categories

Nearly all metallic manufactured products start out as wrought materials, forgings, or castings. Each of these products and their subclassifications will be defined and the usual defects will be described. Wide ranges of quality are available in these raw materials.

Some sort of inspection understanding between the manufacturer and purchaser should exist prior to purchasing. This usually takes the form of an independently written specification, such as those prepared by the American Society for Testing and Materials (ASTM), but may be modified by either party. When such an agreement exists, it becomes a guide for inspection and visual examination.

Wrought materials are the primary products that are produced by hot forming. In this discussion, we will further define them as "steel mill products." Forgings are, in fact, a wrought material and are sometimes produced by a steel mill. They are more commonly the result of a secondary operation producing specific forms that utilize and re-form steel mill wrought products. Castings are forms produced by pouring molten metal into a mold to produce a usable shape. These products are usually only an intermediate stage of manufacture and will be further machined or otherwise fabricated before becoming a usable product.

Wrought Materials

The wrought products of the steel mill fall into the following categories:

Ingots

Ingots are the first product formed when molten metal is poured into large molds. The molds are generally square or rectangular in cross section. The vertical dimension is much greater than the cross-sectional dimension. Ingots may range in size from several hundred pounds to several hundred tons, but are usually in the 10- to 50-ton size. Ingots are almost always reheated and hot worked to form other products. If they are not inspected for defects and conditioned, defects will be carried over to the subsequent product. Strictly speaking, ingots are castings until hot deformed.

Blooms, Slabs, and Billets

Blooms, slabs, and billets are hot formed blocks of steel that are the first stages of re-forming of the large cast steel ingots. Generally, a slab has a rectangular cross section, from 2 to over 20 inches thick, and from 24 to 80 inches wide. A bloom usually has a square cross section, ranging in size from 6 by 6 inches to 24 by 24 inches. A billet has a round-cornered square cross section, but is smaller than a bloom. Billets are generally 2 by 2 inches to 8 by 8 inches. These products are all available in a wide range of lengths. They are usually conditioned at the mill to remove defects. Conditioning takes the form of pickling, grit blasting, scarfing, chiseling, or grinding.

Merchant Bars

Merchant bars are the small rounds, squares, hexagons, and flats in the range of one to two inches in maximum section size. In number of pieces, merchant bars are the most widely used as raw material for manufactured products. They are surpassed in tonnage used by other raw materials.

Plates

Plates are defined as flat, hot-rolled, finished products that conform to the following table:

Width (in.)	Thickness (in.)	Weight (lbs/sq ft)
8 to 48	0.230 min.	9.62 min.
Over 48	0.180 min.	7.53 min.

Plates are usually hot rolled from slabs, but may be rolled directly from ingots. They are usually sheared or torch cut to dimension at the sides and ends, but may be used as rolled.

Structural Shapes

Structural shapes are finished and semifinished products that are hot rolled, generally from reheated blooms. They take the forms of I-beams, channels, rounds, and angles.

Hot Strip Mill Products

These products are basically sheet and strip, with and without coatings such as galvanizing. This material is generally produced in very long lengths and is often coiled. It is usually a semifinished product. The following dimensions for sheet help define the product:

Thickness in Inches	Width in Inches
0.0255 to 0.2030	Up to 3½
0.0344 to 0.2030	3½ to 6
0.0449 to 0.2299	6 to 12

The classifications are general, as other products certainly exist.

Wire

Wire is the result of further hot reducing merchant bars to small diameters. The diameter may be as large as one inch. Wires are more generally under 1/2 inch to about 1/5000 inch in diameter. Although most wire is round in cross section, it can take any one of many other shapes.

Tubular Products

Tubing is available in a large range of sizes, generally from 1/8 inch in inside diameter to 96 inches in outside diameter.

Tubular products are produced by spiral welding, by welding longitudinally after forming, or by piercing and forming (seamless). Tubular products are either finished or semifinished. They are generally round, but may take other shapes.

Defects

Each of the basic steel mill defects will be listed and defined. Except for the ingots, each of these products is a further refinement of other products. For example, wire is made from merchant bars, which are made from blooms, which are made from ingots. The defects developed at each preceding step can be passed on at the next step, if it is not observed and corrected. In addition, each stage of production can produce its own brand of defect, which can also be passed on.

Ingot Defects

Some of the defects found in ingots are categorized as follows:

1. *Cracking.* This may occur longitudinally and on a transverse plane. Longitudinal cracking may indicate excessively high pouring temperatures and coarse large dendrites. These lead to further problems in subsequent mill reductions. Splashing folds in the surface during pouring can produce transverse cracks. Other causes of cracking are worn molds and long delays between pouring and stripping. Such cracking may be observable upon visual examination, but often appears on subsequent rolling operations.
2. *Scabs.* These are surface conditions on ingots caused by the hot molten metal splashing on the mold surface, solidifying and oxidizing such that it will not re-fuse to the ingot. If a scab is not removed by ingot conditioning, it will produce defects in further forming operations.
3. *Pipe.* This is a cavity formed by shrinkage during the last stage of solidification of the molten metal, or by movement prior to solidification (Fig. 32 and 33).
4. *Blowholes.* These are holes produced during the solidification of the ingot, caused by underdeoxidation and by evolved gas which becomes entrapped (Fig. 32).

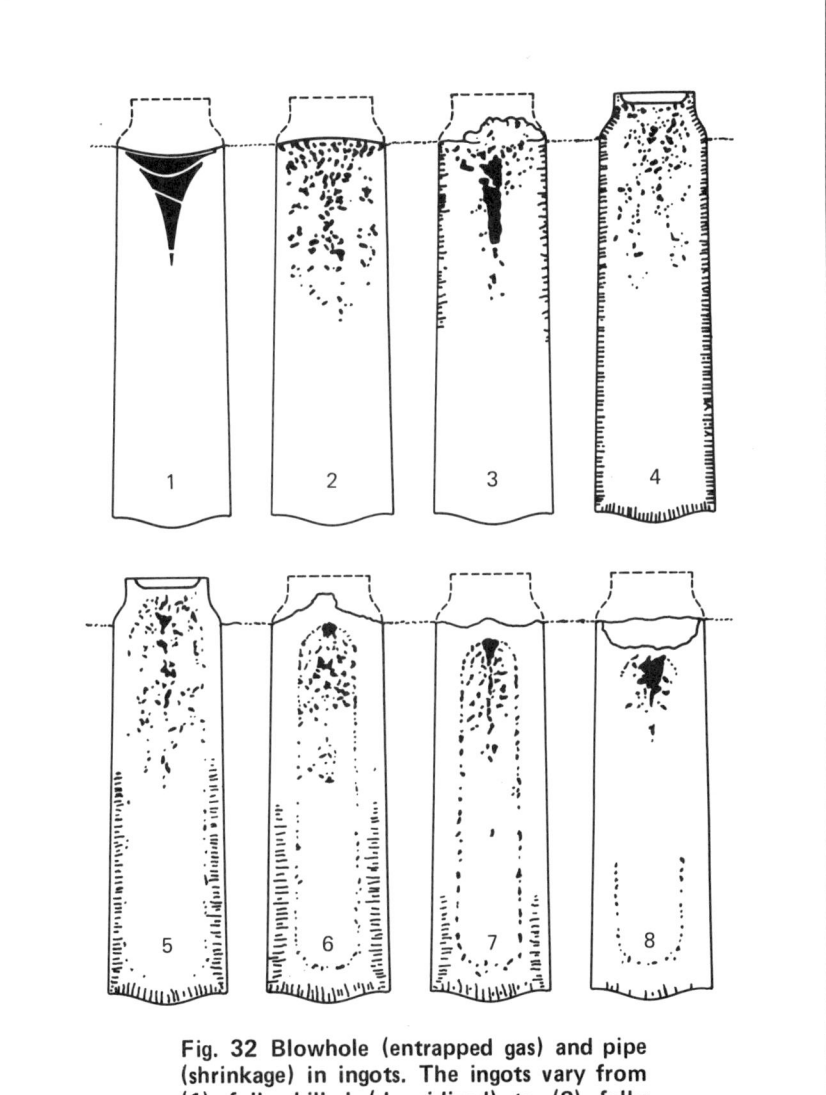

Fig. 32 Blowhole (entrapped gas) and pipe (shrinkage) in ingots. The ingots vary from (1) fully killed (deoxidized) to (8) fully rimmed. (Courtesy U.S. Steel Corp.)

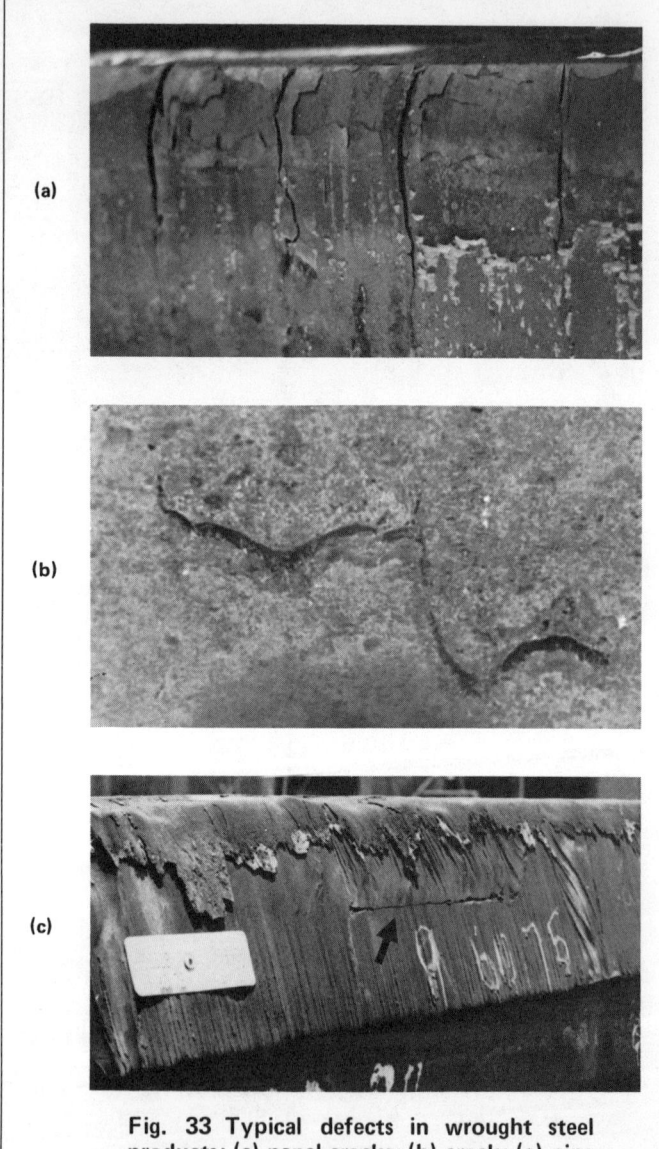

Fig. 33 Typical defects in wrought steel products: (a) panel cracks; (b) crack; (c) pipe (at arrow).

Fig. 7. Temper color chart. (Courtesy Bethlehem Steel Corp.)

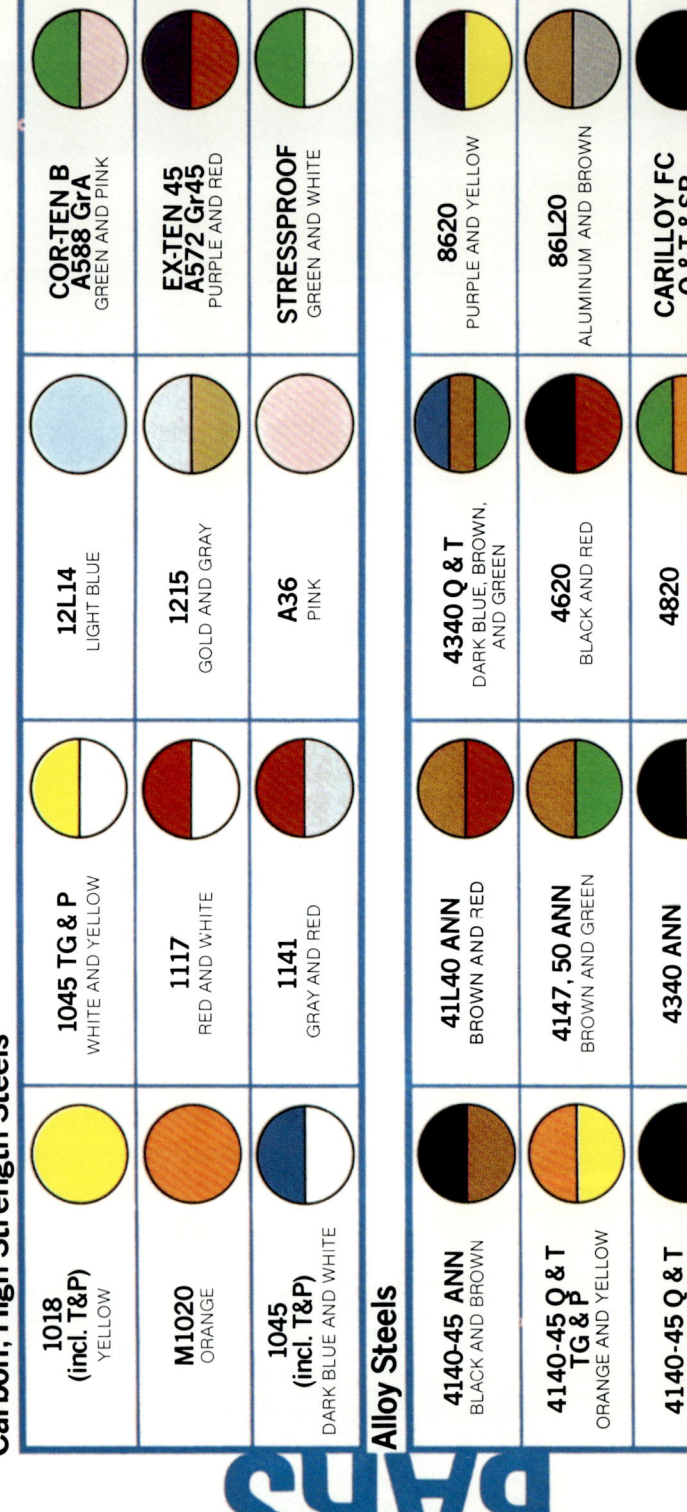

Fig. 38. Typical color code for steel bars and plates. (Courtesy U. S. Steel Corp.)

The seriousness of ingot defects depends upon the ultimate use and type of material. Such defects can be removed by conditioning prior to subsequent hot working if they are likely to affect the ultimate product.

Defects in Blooms, Slabs, and Billets

These fall into the following categories:

1. *Cracks*. These may result from ingot cracks not removed or ingot cracks not observed prior to hot deformation. Cracks can result from too hot or too cold rolling temperatures. They can result from steel of high sulfur content (hot shortness) (Fig. 33 and 36).
2. *Scabs*. These occur during ingot production, but may become visible only upon hot reduction (Fig. 34).
3. *Seams*. These result from elongated ingot cracks, and occur during hot deformation. Transverse ingot cracks produce Y-shaped seams during hot deformation. Other ingot surface defects can produce a whole series of linear grooves or groups of seams (Fig. 34).
4. *Burned steel*. This is severe localized overheating, such as is caused by flame impingement, which produces grain-boundary oxidation and a generally unsalvageable steel (Fig. 35a).
5. *Rolled-in foreign material*. Occasionally, objects such as nuts and bolts come loose from the rolling equipment and are rolled into the product. Scale, slag, and various nonmetallics may also be rolled in as surface defects (Fig. 35b and 36).
6. *Laps*. These are rolling projections that fold over and are rolled into the product, but are not welded due to oxidation.

There are many other defects, some of which involve dimensions, tolerances, and deformations in forming. These are, for the most part, obvious and readily understandable.

Although not all of these defects can be removed completely, the degree of removal is dictated by the final product. Cost of removal is also an important factor. Pickling after hot forming (rolling) makes defects more easily discoverable. Steels to be heat treated, particularly those of high hardenability, are susceptible to quench cracking. This will more readily occur in the presence of surface defects, particularly cracks and seams. Careful visual

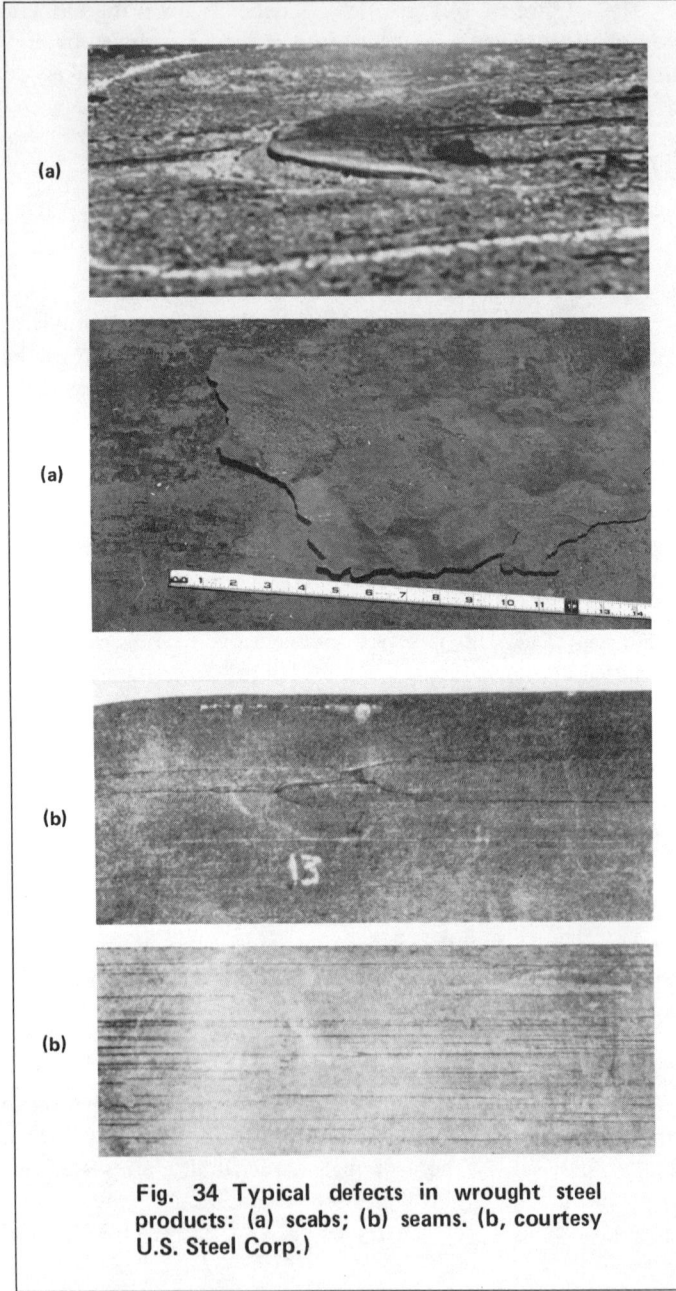

Fig. 34 Typical defects in wrought steel products: (a) scabs; (b) seams. (b, courtesy U.S. Steel Corp.)

Fig. 35 Typical defects in wrought steel products: (a) burned steel; (b) rolled-in nonmetallics.

examination using good lighting will reveal most of the important defects.

Specifications should be prepared governing acceptable defects, removable defects, and rejectable defects. If possible, photographic standards should be prepared. Re-inspection after defect removal is important to ensure that the removal process did not produce additional defects or open up new ones.

Inspection of Plates

Since plates are much reduced in section by hot forming and have large surface areas, defects become magnified and are much

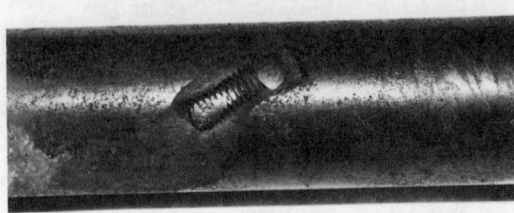

Fig. 36 Typical defects in wrought steel products: foreign material rolled into the product.

easier to identify. Visual inspection of plates involves both surfaces and all edges. The cracks, scabs, seams, and laps previously described also carry over to plate form. Plate examination requires adequate lighting. The following other defects become apparent:

1. *Blisters.* These are defects in plate, on or near the surface, caused by the expansion of gas pockets beneath the surface (Fig. 37). Blisters in particular require low-incidence side lighting to see.
2. *Laminations.* These are visible at the sheared edges and ends of the plate. They are a mid-wall plane of separation, weakness, or entrapped foreign material. They are aligned parallel with the worked surface of the material. They may be the result of pipe, blisters, seams, inclusions, or segregation.
3. *Scale.* Since the surface appearance is often of some importance, rolled-in scale and foreign material should be evaluated for degree of acceptability.

Examination of Raw Materials and Semifinished Products / 59

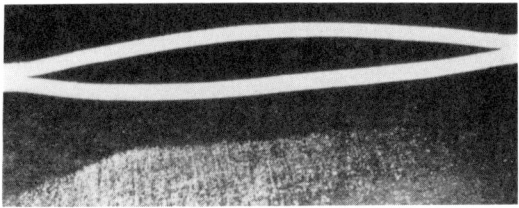

Blister (Sectioned)

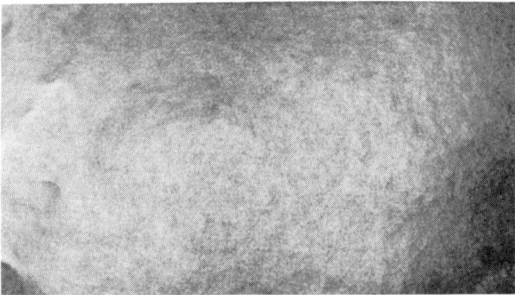

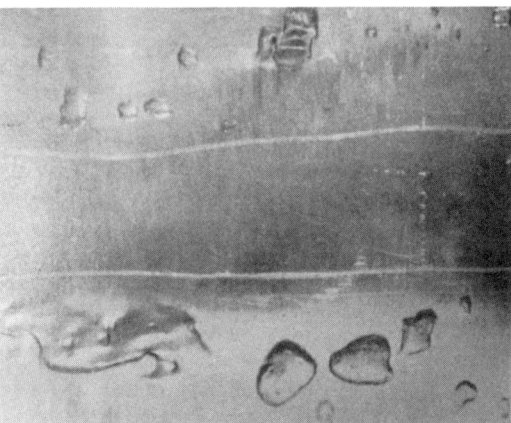

Fig. 37 Typical defects in wrought steel products: blisters. (Top photos courtesy American Petroleum Institute; bottom photo courtesy American Rolling Mills)

4. *Hard spots.* These may develop during plate rolling due to isolated water leaks impinging on and causing rapid cooling of a particular spot. Discoloration patterns may occur, but are difficult to evaluate by visual examination at this stage.

Structural Shapes

Defects in structural shapes are similar to those already described in detail. Pipe, blisters, scabby surfaces, cracks, and surface conditions are among the more important. Since these are often final products, inspection is of considerable importance. Dimensional measurements should be made and the product checked for straightness. Markings such as the mill heat number (Fig. 1) can be observed and used to tie the material to the proper mill test reports.

Inspection of Merchant Bars

Carbon and alloy steel merchant bars suffer from many of the defects previously described. These materials are often heat treated, which may introduce such visual defects as quench cracking, distortion, and scaling. *The Making, Shaping, and Treating of Steel*, published by the United States Steel Corp., describes the following as being the most regularly occurring mill defects in merchant bars:

1. *Fins and overfills*—Protrusions formed when the section is too large for the pass it is entering, or when proper allowance has not been made for lateral spreading in the rolls. Overfills are broad and less sharp than fins. As a rule, overfills occur more frequently than fins and in many cases are associated on the same bar with underfills.
2. *Underfills*—The reverse of overfills; that is, they are areas in which the section is incompletely filled. They are formed by permitting the bar to be rolled scant in certain dimensions. Underfills appear most frequently on rounds and channels.
3. *Slivers*—Loose or torn segments of steel rolled into the surface of the bar. They may be caused by a bar shearing against a guide or collar, incorrect entry into a closed pass, or a tear from other mechanical causes. Sometimes slivers are present in the billet and are carried through to the hot-rolled bar or shape.

4. *Laps*—A rolled-over condition caused by a bar having been given a pass through the rolls after a sharp overfill or fin has been formed, causing the protrusion to be rolled into the surface of the product.
5. *Seams*—Crevices in the steel that have been closed, but are not welded. This type of defect is very difficult to detect on certain types of steel products. Seams are caused by blowholes and cracks in the original ingot, or by faulty methods of rolling in both semifinishing and finishing mills (Fig. 34).
6. *Fire cracks and roll marks*—Impressions in the product, of varying degree and pattern, caused by mill rolls becoming overheated, and cracking or spalling.
7. *Scratches*—Long nicks or indentations in the product caused by the surface or surfaces of the bar rubbing against sharp or pointed objects such as guides on the mill, chutes, "dead" conveyor rolls, chain hoists, or other mechanical equipment.
8. *Rolled-in scale*—A defect in the surface caused by scale, formed during a previous heating, which has failed to be eliminated during the rolling operations. It is one of the most prevalent surface defects.
9. *Buckle and kink*—A corrugated or wrinkled surface condition caused either by worn-out pinions on a roll stand or uneven cooling beds. A buckle is an up-and-down wrinkle; a kink is a side wrinkle.
10. *Burned steel*—Appears as a rough area with checked or serrated edges (Fig. 35a). Burning is caused by exposure of the steel to an excessive temperature. Burned steel is always scrapped.
11. *Camber*—The deviation of the side edge of a bar from a straight line. It is caused by improper heating of the billet, uneven dimensions causing differential expansion or contraction, or improper alignment on the hot beds.
12. *Hook*—A short bend or curvature caused either by improperly adjusted delivery guides or by any obstruction that momentarily halts the forward motion of the bar from one roll stand to another.
13. *Pipe*—A steelmaking defect carried through from the ingot. The presence of pipe is detected as a cavity located in the center of an end surface.
14. *Shear distortion*—A mashed or deformed end on a bar caused by defective or improperly adjusted shearing equipment.

15. *Twist*—A condition wherein the ends of a bar have been forced to rotate in relatively opposite directions about its longitudinal axis. It may be caused by excessive draft, faulty setting of delivery guides, or lack of uniform temperature in the bar.

Larger merchant bars are marked with heat numbers; smaller bars are bundled and tagged with heat numbers. These numbers can tie them with mill heat reports. Manufacturers may color code them by types, painting different colors on the ends of the bars. When a section of bar that includes heat identification is removed, repainting of the end will avoid loss of identification. A typical color code is shown in Fig. 38 (see insert between pages 54 and 55).

Defects in Wire Products

Wire defects are often revealed during manufacture. Wire is produced by drawing rod through a series of reducing dies. Since this puts high loads on the product, major defects will cause breakage during drawing.

Wire products have many of the carry-over defects produced in earlier raw material reduction stages. The small size of the wire makes defects difficult to observe. The general problems are improper size and shape, which measurements reveal. Internal defects—such as pipe, seams, and segregation—are generally too minute to observe visually and require destructive examination. Surface defects include scratches, slivers, and seams. The scratches occur at the dies due to wear, poor lubrication, or entrapped foreign objects. Slivers produce pointed projections, which rise from the surface of the wire. Seams are longitudinal cracks and may be too small to detect visually.

Defects in Tubular Products

Welded tubing is made from hot-rolled skelp. Skelp is a strip of metal produced to a suitable thickness, width, and edge configuration from which pipe or tubing is made. The defects that occur in skelp are those already described for plates. When the flat skelp is deformed to produce the contour of the pipe and tubing, flat spots may be observed. These are indicative of hard spots in the skelp and are undesirable from the standpoint of dimensions and corrosion. They may also crack. Since the forming operation involves exceeding the yield strength of the material, defects that had been missed during previous inspection can be observed.

Submerged-arc welding, butt welding, or electric resistance welding, used to seal the formed edges, produce their own defects. (Weld defects are discussed in detail in Chapter 4.) Hydrostatic pressure testing of pipe and tubing reveals most weld defects. Since the defects are often very small and tight, materials to reduce the surface tension of the water and dyes to make leaks more visible may be used. Whenever pipe is welded, arc burns may occur. These are caused by arcing between pipe and electrodes or controls. They may produce localized melting or hard, crack-sensitive heat-affected zones. If not severe, they can be ground out.

Denting and gouging often occur in tubing due to mechanical effects of handling. All pipe and tubing should be checked for eccentricity. This can be out-of-roundness or variation in wall thickness around the circumference. Measurements of inside and outside diameters around the periphery will quickly reveal any such variation.

Seamless pipe and tubing are also susceptible to the many mill-type defects previously described. Seamless material should be carefully examined internally and externally because the manufacturing process is prone to produce internal defects. Such defects are plug scars, which are grooves caused by adhering metal on the forming die or plug. Roll marks are forming defects on the surface. Slugs are foreign material imbedded into the surface, but not fused. Other defects found in seamless pipe and tubing are blisters, gouges, laminations, laps, pits, scabs, and seams.

Defects in hot strip mill products are similar to those in steel plate. Since these products are thinner, surface conditions are more important. Defects not previously revealed may be apparent. Flatness or levelness may be critical. Surface conditions relate to material defects or rolled-in scale. They may also occur because the material is not properly heated, rolled, pickled, or processed. Another cause of surface defects is the condition of the rolls themselves.

Forgings

Forgings are of various types. The most common are open-die or hammer forgings. These are produced by heating a piece of forgeable material and hammering it to form, with or without a die cavity. Horseshoes fall in this category as do many large shapes such as spindles, rotors, gears, and hubs. Closed-die forging shapes metal completely within the walls of two mating dies. This process is principally confined to smaller parts and produces more complex shapes with closer tolerances than open-die forging.

Upset forging or hot upset forging involves gripping one end of a heated bar or tube and applying pressure to the end protruding, to upset and re-form that section only. For example, the heads on bolts are produced that way, as are the enlarged ends of pipe that are internally threaded. There are other varieties of forging, but they are either similar to these three in the area of defects, or not produced in large volume.

Defects in Forgings

Forging defects fall into two categories: those resulting from defects in the material being forged, and those resulting from the forging process (Fig. 39). Defects in the material have been discussed under the various sections on wrought materials. Producing a forging involves reheating an ingot, a bloom, a billet, or a bar, and hot deforming it. Hot deforming accentuates defects present in that material. Such defects can also lead to the formation of new defects.

When an ingot is the starting point, as for many large forgings, the segregation of the cast ingot produces certain problems. If the segregation is not carefully and uniformly broken down during forging, cracking can occur along these planes of segregation. Nonuniformity in response to subsequent heat treatment can also result.

Centerline shrinkage or internal weakness can result in rupture during forging. This is revealed at the time of occurrence or after rough machining. It takes the form of radial cracks called bursts, originating at the center of the forging and radiating in various directions.

If hydrogen is present in the steel, internal flakes may occur upon forging. These again are often revealed upon machining.

The internal surfaces of such flakes are often bright due to the reducing action of the hydrogen (Fig. 39b).

Nonmetallic inclusions are undissolved materials in the original cast ingot (Fig. 39f). In excess, these induce cracking and other problems during forging. They contribute significantly to service fractures. Visually, they are only observable if they open up during forging or upon machining.

Forging defects resulting from weld repairs on raw material may take the form of slag, gas pockets, unfused electrodes, or hard and soft zones. All may be visually revealed on examination of the machined surfaces after forging.

Uneven solidification or cooling rates at the ingot surfaces may produce a condition called shelf, which is a layer or skin effect (Fig. 39a).

Surface laps occur when hot metal is folded over during forging, and oxidation prevents re-fusing. These are the same as described for steel mill defects.

Seams are those described in wrought material defects, as are slivers and rolled-in scale.

Ferrite fingers are light-appearing rivers on machined surfaces (Fig. 39c). They result from decarburization that is forged in or cracks that weld together. They may have entrapped oxides and are softer than the surrounding material. This makes them show up readily on a machined surface because they machine differently.

Fins and overfills (Fig. 39e) occur as the result of improper working during reduction.

Cracking may result from shearing the flash from the edges of the forgings. These cracks are diagonal and occur if the trimming dies are improper or in bad condition. Improperly designed dies can produce various degrees of defectiveness if the material cannot adequately fill the die.

Heat treatment after forging can also result in cracking, burning, decarburization, or excessive scaling. This may or may not be visually discoverable. Heating and cooling rates, as well as time and temperature, are potential causes of defectiveness. Cracks open above 1050 F will scale appreciably in the presence of air. Below this temperature, slight discoloration will occur on the fracture surface of the separated crack. Temper color charts such as that in Fig. 7 (see insert between pages 54 and 55)

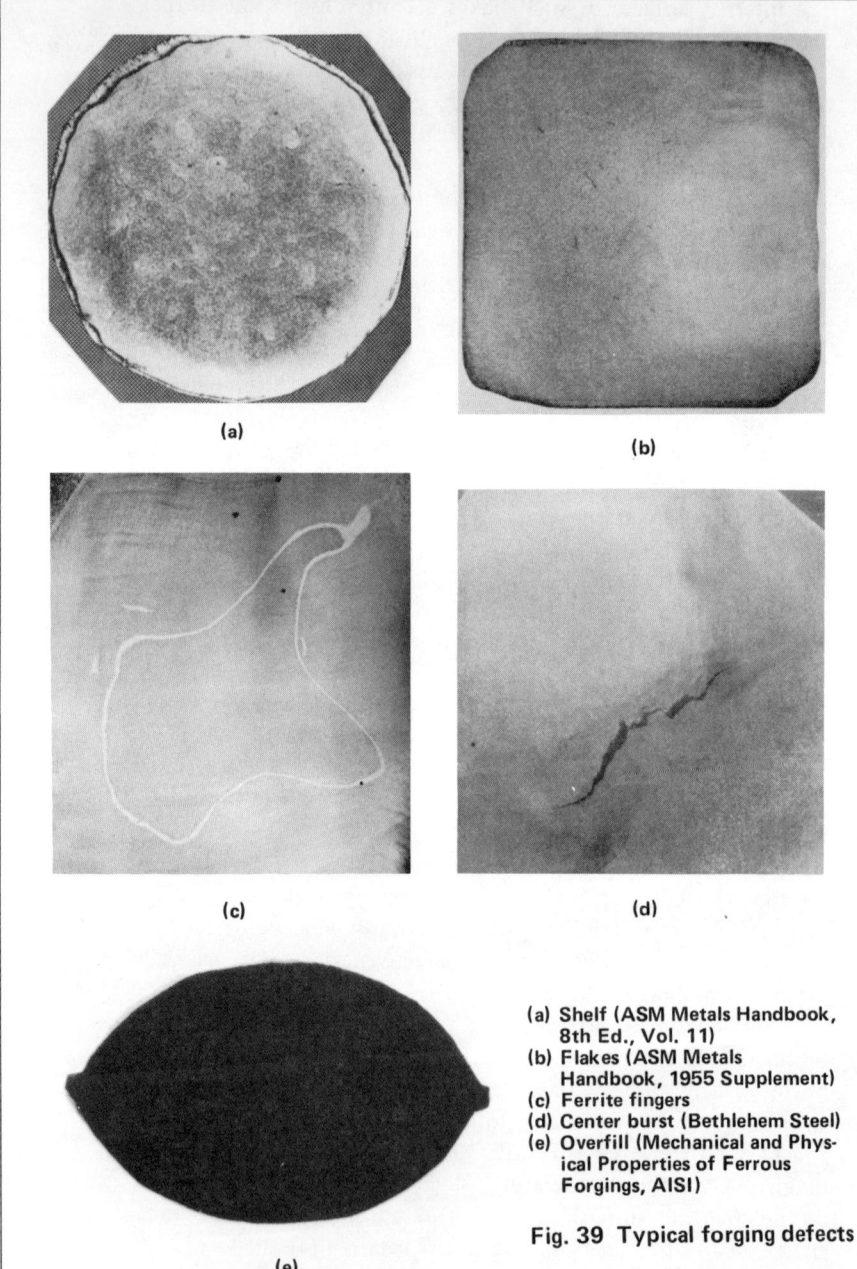

(a) Shelf (ASM Metals Handbook, 8th Ed., Vol. 11)
(b) Flakes (ASM Metals Handbook, 1955 Supplement)
(c) Ferrite fingers
(d) Center burst (Bethlehem Steel)
(e) Overfill (Mechanical and Physical Properties of Ferrous Forgings, AISI)

Fig. 39 Typical forging defects

(f)

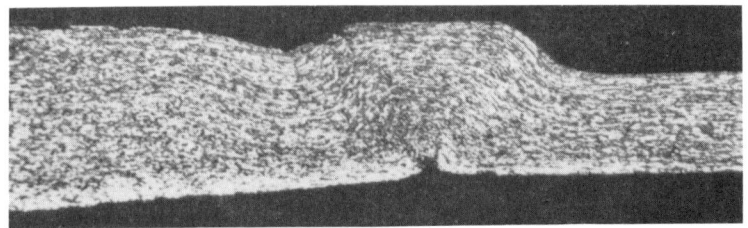
(g)

(f) Nonmetallic inclusions
(g) Wrinkle (American Petroleum Institute Bulletin 5T1)
(h) Underfill (American Petroleum Institute Bulletin 5T1)

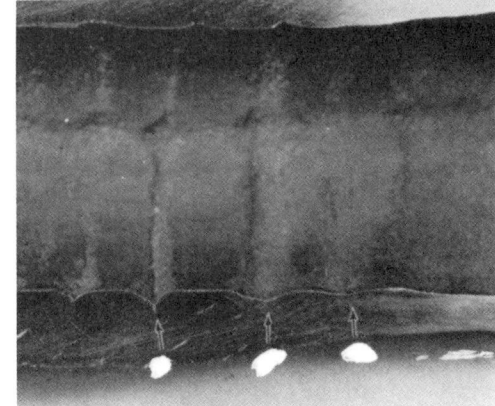
(h)

reveal approximately the temperature the crack has been exposed to and may help determine when it occurred.

Upset forgings may produce a depression on one or more surfaces. This is called underfill and is caused by insufficient flow of metal (Fig. 39h). Wrinkles occur upon upset forging due to buckling-type laps (Fig. 39g).

Swabs used to clean and lubricate dies may lose segments that can then be actually forged into the surface of the part. This will also occur for any nonmetallic material left on the die, forging material, or hammer. In closed-die forgings, offset of the lower and upper parts indicates improper die match.

The Drop Forging Association publishes a compilation of forging industry trademarks in use. Visual examination will reveal such markings, and the manufacturer can be identified. These markings are similar to those seen on the bolt heads in Fig. 5.

Castings

Castings, like forgings, are of several varieties, which are related to the type of mold the molten material is poured into. Sand mold castings account for the largest tonnage. Although castings are made from many materials, this discussion will be limited generally to the visual examination of steel sand castings. Many of the problems involved are applicable to the other casting methods.

Casting Defects

Casting defects fall into various categories:

1. *Entrapped gas, porosity, and blisters.* Visual examination reveals evidence of gas entrapped if it reaches the surface of the casting. It may take the form of large pockets, pinhole porosity, or a blister-like thin skin of metal covering such a gas pocket (Fig. 40). It is caused by many factors, including improper molding, melting practice, pouring practice, and moisture.
2. *Shrinkage.* This is a depression in a casting wall, which often has a jagged dendritic appearance. It is caused by the contraction of metal during solidification, generally because of

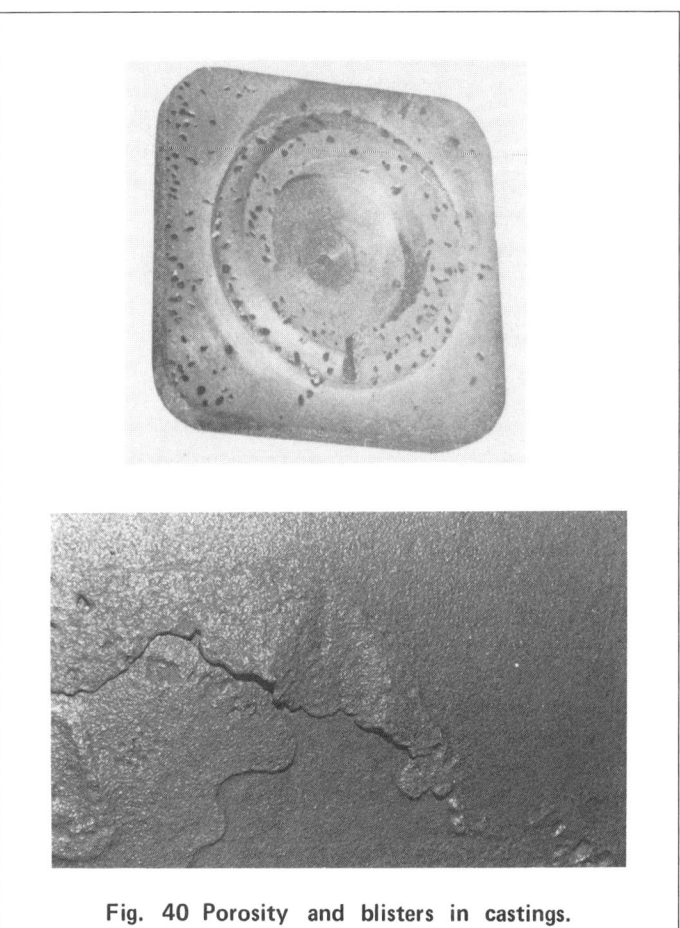

Fig. 40 Porosity and blisters in castings. (Courtesy American Foundrymen's Society)

use of improper risers and improper pouring procedures (Fig. 41).
3. *Hot and cold tearing and cracking.* These are forms of cracks that may or may not surface. They are caused by differing reactions. Hot tears (Fig. 42) are generally more widely opened and result from contraction during solidification if movement is restricted. They are heat discolored, oxidized,

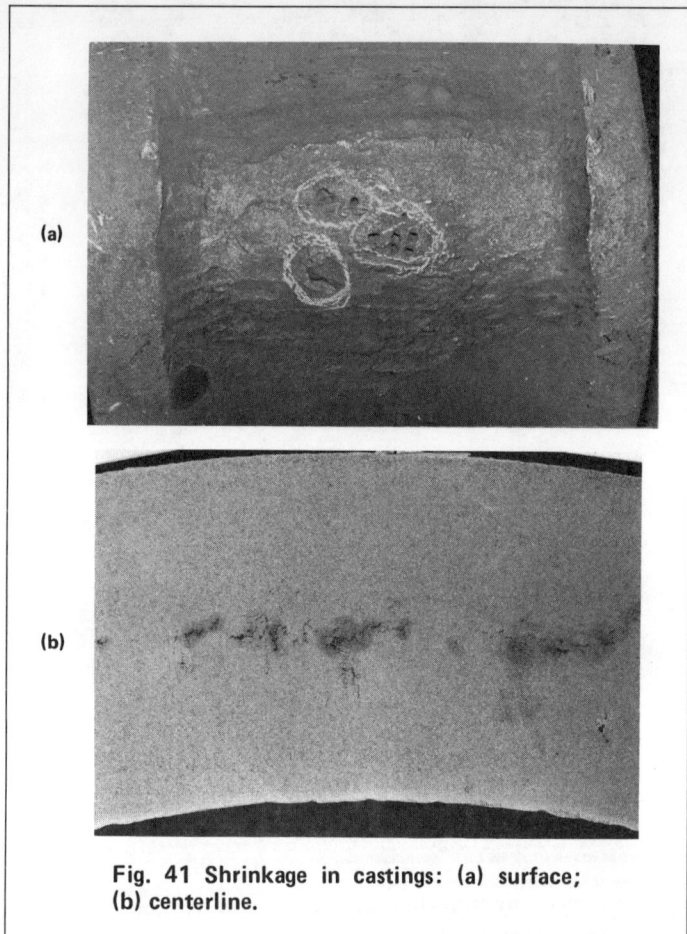

Fig. 41 Shrinkage in castings: (a) surface; (b) centerline.

and/or decarburized. Hot cracks are caused by the stresses after solidification and during cooling. They are generally tighter cracks than hot tears and may be heat discolored, but not scaled or decarburized. Cold cracking results from handling stresses after cooling and produces bright fracture surfaces. Heat treatment after casting can produce cracking that is more similar to the hot cracking.

4. *Surface conditions caused by contamination.* These have various names, such as:

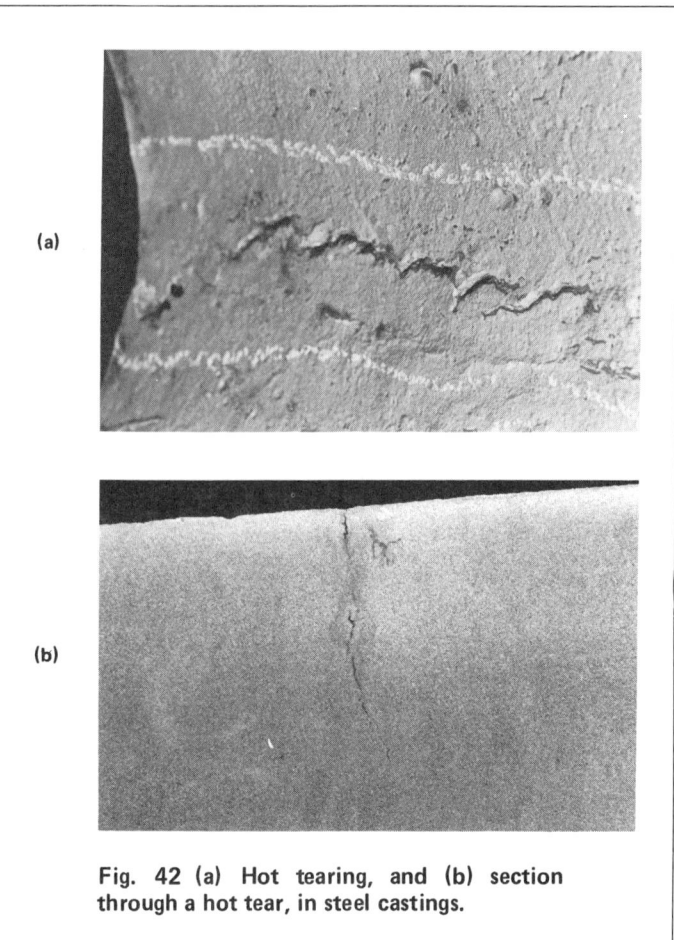

Fig. 42 (a) Hot tearing, and (b) section through a hot tear, in steel castings.

a. *Inclusion*–Sand, dirt, or other foreign material entrapped at the casting surface (may be ground out if not too deep).
b. *Slag*–Surface entrapment of the dross, which generally floats to the top of the ladle of molten metal (Fig. 43a).
c. *Fusion*–Melting or fusing of sand at the cast surface producing a glossy effect.
d. *Rat or sticker*–Part of the mold stuck to the casting surface.
e. *Crushes, pushups, and clamp-offs*–If the mold sand moves

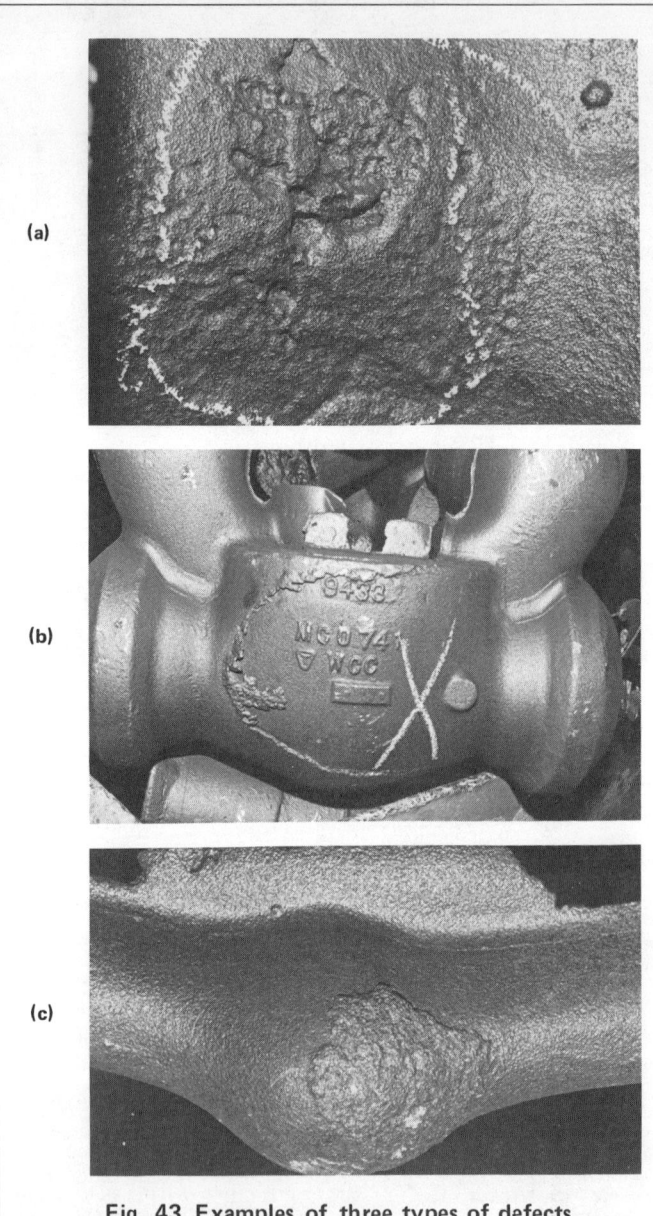

Fig. 43 Examples of three types of defects in steel castings: (a) a slag inclusion; (b) a pushup; (c) an erosion scab.

due to nonuniform pressure caused by any of these, the casting surface will have sunken areas (Fig. 43b).
f. *Erosion scabs*—If the mold sand is washed away by excessive motion of the molten metal, a mass consisting of a mixture of solidified sand and metal will result (Fig. 43c).
5. *Incomplete castings.* There are various reasons for castings to fail to completely fill out the mold:
    a. *Misrun or short pour*—Failure of molten metal to completely fill the mold cavity (Fig. 44a).
    b. *Cold shut*—When two streams of metal within a casting are being poured and fail to fuse properly, a round-edged crack or seam results.
    c. *Scar*—Result of entrapped gas preventing complete filling of the mold.

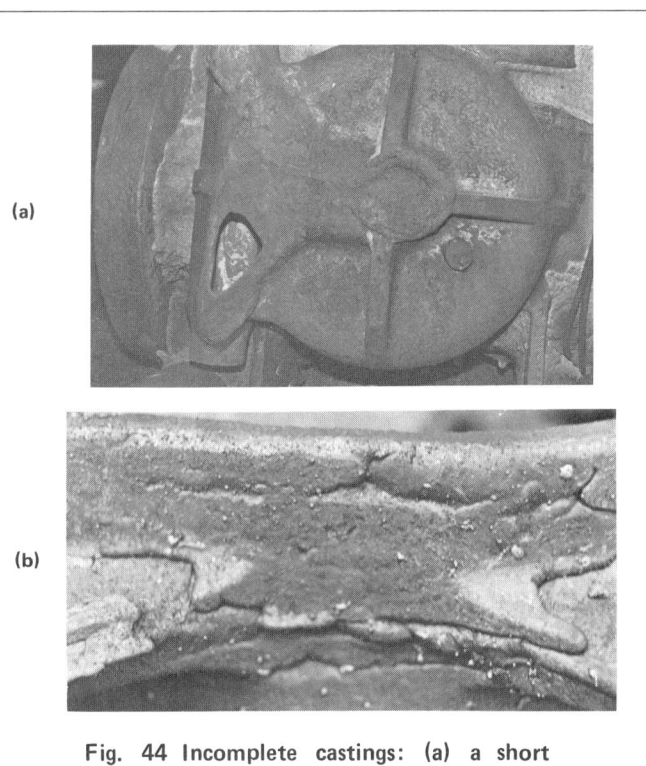

Fig. 44 Incomplete castings: (a) a short pour; (b) a runout.

d. *Plate*—Material formed when metal fills a scar, but does not fuse.
e. *Runouts*—Mold leak allowing molten metal to run outside the mold while being poured (Fig. 44b).
f. *Bleeder*—Molten metal running out of the mold after pouring has been completed.

6. *Other surface conditions*:
   a. *Buckles*—Dents in the casting surface due to sand expanding.
   b. *Rattails*—Irregular lines caused by minor buckles.
   c. *Pull downs*—Buckles in the upper part of the casting.
   d. *Scabs*—Thin layers of steel separated from the casting by a thin layer of mold sand broken loose. They are joined to the casting at some point (Fig. 45a).
   e. *Wrinkles*—Smooth depressions resulting from irregularities in pouring practice (Fig. 45b).
   f. *Veins*—Raised narrow linear ridges caused by mold or core cracking (Fig. 45c).

7. *Misshapen castings*:
   a. *Warpage*—Cooling deformation.
   b. *Shifts*—Mismatches of mold halves (Fig. 46a) or core movements that create dimensional problems (Fig. 46b).
   c. *Cuts and washes*—Changes in dimensions due to mold or core erosion by either molten metal or gases.
   d. *Swell*—An increase in section size due to pressure on the mold wall.
   e. *Fin*—A thin metal projection.
   f. *Sag*—A thinning of the section due to mold distortion.
   g. *Drop*—A casting defect caused by sand from the upper portion of the mold depositing on the lower portion.

Many castings are heat treated after being cast, and defects from heat treating may resemble some of the casting defects described.

Flame cutting of gates and risers, grinding with abrasives, and cutting and gouging with torches for repairs are all potential sources of defectiveness. These defects take the form of burning, hard zones, or cracking.

When defects are too deep to remove, they must be repair welded. Such repairing is an additional source of possible defects.

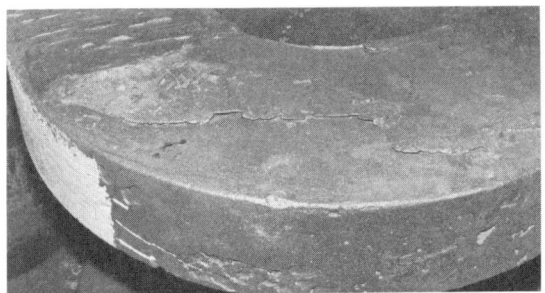

(a)

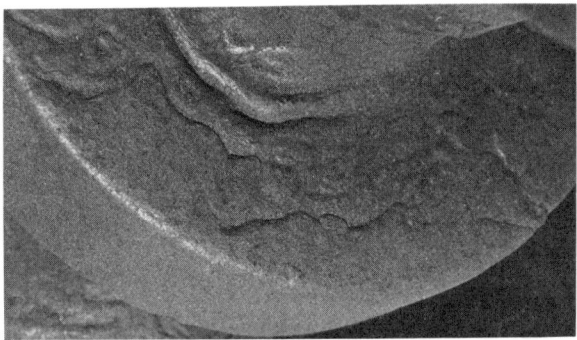

(b)

(c)

Fig. 45 Casting surface defects: (a) scab; (b) wrinkles; (c) veins (at arrow).

Fig. 46 Misshapen castings: (a) mismatch; (b) mold shift.

(Examination of welds for defects is discussed in detail in Chapter 4.) The presence of repair welding, unless otherwise specified, is not in itself cause for rejection.

There are different markings found on the surfaces of castings that may be significant to the examination. Foundry marks are symbols usually cast on the part that identify the maker. The material specification may be cast on the surface. This is usually an ASTM specification, but may only identify a particular grade.

Fig. 47 Identification markings on a steel casting: (A) pattern number; (B) foundry symbol; (C) heat number (at arrow); (D) material type.

For example, ASTM Specification A216 Grade WCB may only show the symbols "WCB" on the casting. The heat number identifying the batch of steel which ties the casting to a mill test report may be a cast number or may be a steel-stencil stamp on the surface. It may be randomly stamped, or there may be a special boss where it is located (Fig. 47).

CHAPTER 4

Fabrication

In the past, riveting and bolting were the more common methods of attaching metal parts to one another. Although these methods are still in use, welding, brazing, and soldering have increasingly become more common methods in fabrication.

Welding

Welding is done by many processes, but all involve different methods of joining metals by heating them to the melting point and fusing them. This can mean melting the metals to be joined (similar or dissimilar) or the addition of another metal (welding rod, wire, or powder). The process is accomplished by vastly differing methods, the most common of which is arc welding. Other processes are gas welding, friction welding, electron-beam welding, resistance welding, forge welding, spot welding, thermite welding, and many sub-varieties. The principal differences in the types involve the methods of heating and fusing.

Welding, especially arc welding, is part science and part art. Virtually anyone can be trained to weld, but a good welder is both artist and craftsman. This is one of the reasons inspection of welds is so important.

Visual Examination Prior to Welding

Visual examination prior to welding can reveal many potential problems. Dimensional measurements should be checked for accuracy. If parts are weld beveled, check the bevel. Does it conform to the drawings and specifications? (Drawings and specifications are the rules of the inspector.) The fit-up should be checked: do the parts align properly? Is the specified welding process being used? If stick electrodes are used, are they kept in a drying oven to avoid moisture? (Damp rods may produce hydrogen problems and cracking.) Is the rod of the type and size specified? Often root passes are required to be made with a smaller-diameter rod than the capping passes. This provides good root penetration.

Parts to be welded should be clean and free of oil and moisture (moisture may condense, particularly early in the morning in humid areas). Oily compounds result from machining, such as beveling. Oil and moisture produce hydrogen, which causes underbead cracking. They also produce gases, which become entrapped in the weld metal causing a series of pinholes, often down the centerline of the weld. When multipass welds are required, excess slag should be cleaned between passes to avoid entrapment.

If preheating and interpass temperatures are specified, these should be observed and temperatures should not be allowed to fall below these temperatures. Cooling rates after welding should be ascertained. Some welds must be slow cooled to avoid cracking and excessive distortion. This may be accomplished either in a furnace or by covering with insulating material. Other weldments must be heat treated after welding. In some cases, they must be transferred immediately to the heat treating furnace before cooling completely.

Although careful visual inspection during welding is desirable for production of good welds, more often the weld is inspected long after it is made. Such visual inspection can reveal much about the weld quality even though the internal structure cannot be seen. The first thing to observe is shape. Is the part within specified dimensional tolerances? Has welding caused warping and distortion? Warping and distortion can not only prevent normal usage, but also be indicative of high locked-up stresses. The second thing to observe is the outer surface appearance of the weld. A well-made weld should have a good, attractive appearance. Figure 48 shows the effect of welding variables on weld appearance.

With proper amperage and welding speed, the weld bead has an even contour with regular ripples and complete penetration. This weld has an attractive appearance (Fig. 48f).

If the amperage is too low, the bead ripples are uneven and have an unattractive appearance (Fig. 48b). The bead tends to be narrow and build-up is high. Undercutting occurs, producing stress concentration notches at the edges of the weld. The weld does not penetrate completely to the root or backside of the weld.

If the amperage is too high, the weld bead ripples are elongated and weld splatter is produced (Fig. 48d). In this case the penetration of the weld metal is excessive, producing a large lip-like protrusion at the underside.

If the welding speed is too slow, the weld metal builds up on the surface, producing a high contour (Fig. 48a). The penetration is poor, resulting in stress concentrations and wall thinning.

High-speed welding produces narrow, uneven ripples at the bead surface and undercutting (Fig. 48c). The penetration is poor.

Another variable is the arc length. Even if the amperage and speed are proper, if the arc is too long, porosity and inclusions will result. With a long arc the penetration is irregular (Fig. 48e). Low voltage and high voltage act much like low and high amperage.

During welding, the smoothness of the electrode melting and steadiness of the arc are indications of the likelihood of a good, sound weld.

Not all welding defects are revealed by visual examination, but those that are not are often suspect, simply by the general characteristics of the weld. The principal welding defects are as follows:

1. *Incomplete fusion.* This occurs either when the molten weld metal fails to fuse to the base metal or when one pass fails to fuse to another (Fig. 49). It can be caused by:
    a. Failure to produce enough heat to cause melting.
    b. The presence of foreign material (improper cleaning and fluxing).

82 / **Inspection of Metals: Visual Examination**

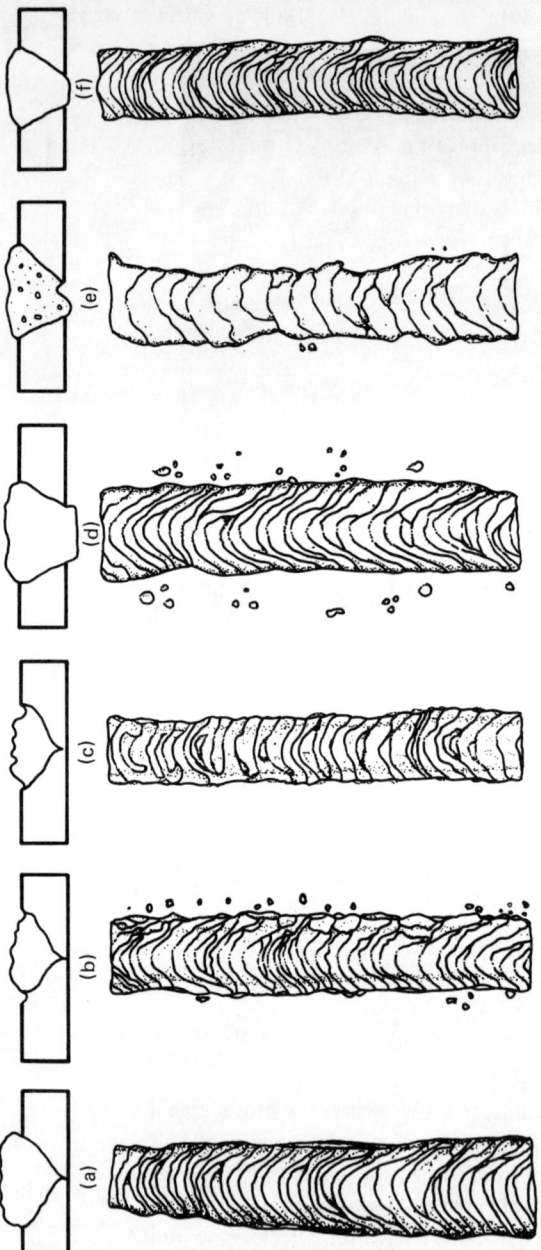

(a) Amperage is normal but welding speed is slow. The weld bead builds up high but the penetration is poor.
(b) The weld speed is correct but the amperage is low. The bead ripples are uneven, there is undercutting, and penetration is poor.
(c) The amperage is normal but the welding speed is too rapid. The ripples in the bead surface are uneven, there is undercutting, and penetration is poor.
(d) The speed is normal but the amperage is high. The bead ripples are elongated, there is weld splatter and excessive penetration.
(e) The amperage is normal but the arc is long. There is uneven penetration, porosity, and entrapped nonmetallics.
(f) All conditions are normal. The penetration is good and the bead contour is uniform.

Fig. 48 Effect of welding variables on weld appearance.

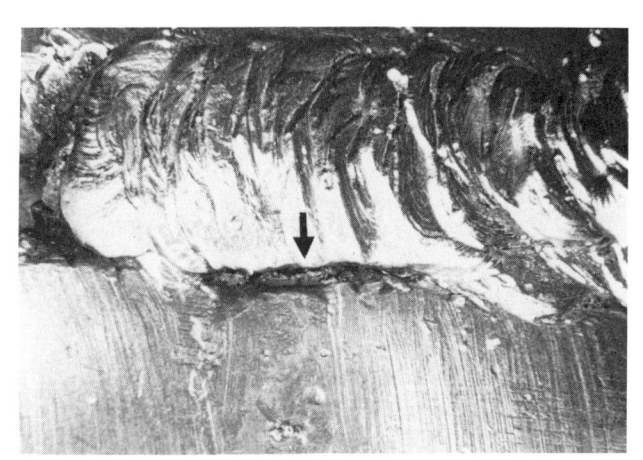

Fig. 49 Incomplete fusion.

    c. Welding of materials that are incompatible. This is not a common defect in welding and is only discoverable visually if it is severe and if it surfaces.

2. *Insufficient penetration.* This is the condition that occurs when insufficient melting and fusion occur at the root or underside of the weld. It produces a depression or notch effect. If severe, it reduces the wall section. More important, it acts as a stress concentration point. It can result in weld failure without appreciable deformation. Incomplete penetration is caused by:
   a. Improper joint design, such as groove shape and root opening.
   b. Use of improper electrode size (often a smaller-diameter electrode is used for the root pass).

   Insufficient penetration is readily revealed by visual examination, if the underside of the weld is visible (Fig. 50).

3. *Undercutting.* This is the groove or notch that is produced parallel to one or both edges of the top surface of the weld bead (Fig. 51). It is caused by:
   a. Operator technique.
   b. Use of certain electrodes.

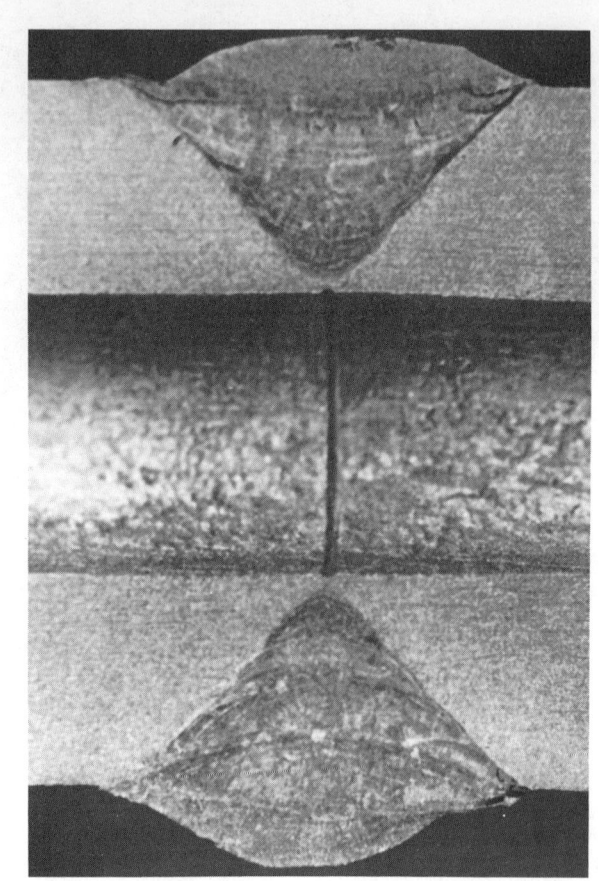

Fig. 50 Insufficient penetration.

c. Welding variables, such as high amperage or voltage, a long arc, low amperage or voltage, and high welding speeds.

Undercutting produces the same outer surface stress concentration effect as insufficient penetration does at the root. Stresses are generally higher at the surface, so the effect may be severe. In addition, the wall thickness is reduced. Any stress concentration area, such as this, is particularly undesirable where cyclic stress is involved. Fatigue failures will occur if these stresses are high enough. Undercutting is always detectable by simple visual examination. It can be repaired by use of an additional welding pass or passes in the grooved area. Small repairs, less than an inch long, should be avoided.

4. *Slag*. This is the term used for the nonmetallic extraneous material that becomes entrapped in the weld metal at the time of solidification (Fig. 52). It is generally metallic oxides and flux materials, in the form of globular or elongated masses. The size ranges from pinpoint to perhaps as large as a dime. All welds have some entrapped slag, and when the quantity and size are small, it causes no problem. Slag in welds is caused by:
   a. Low welding temperatures which cause rapid cooling, not allowing the slag to float to the surface.
   b. High welding temperatures which generate excessive oxides.
   c. Narrow, deep weld grooves.
   d. Rapid cooling due to large masses of cold, adjacent metal and no preheat.

   Only slag exposed at the surface can be observed visually.

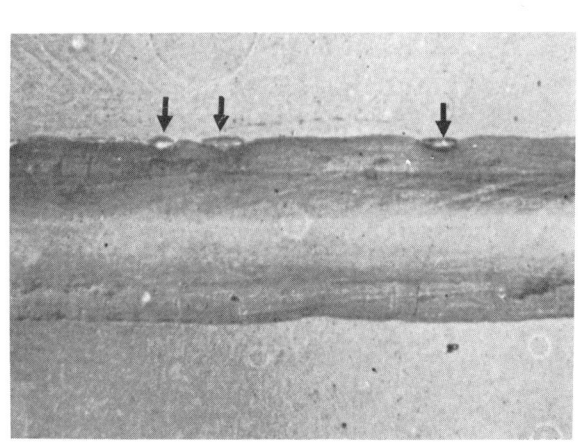

Fig. 51 Undercutting (at arrows).

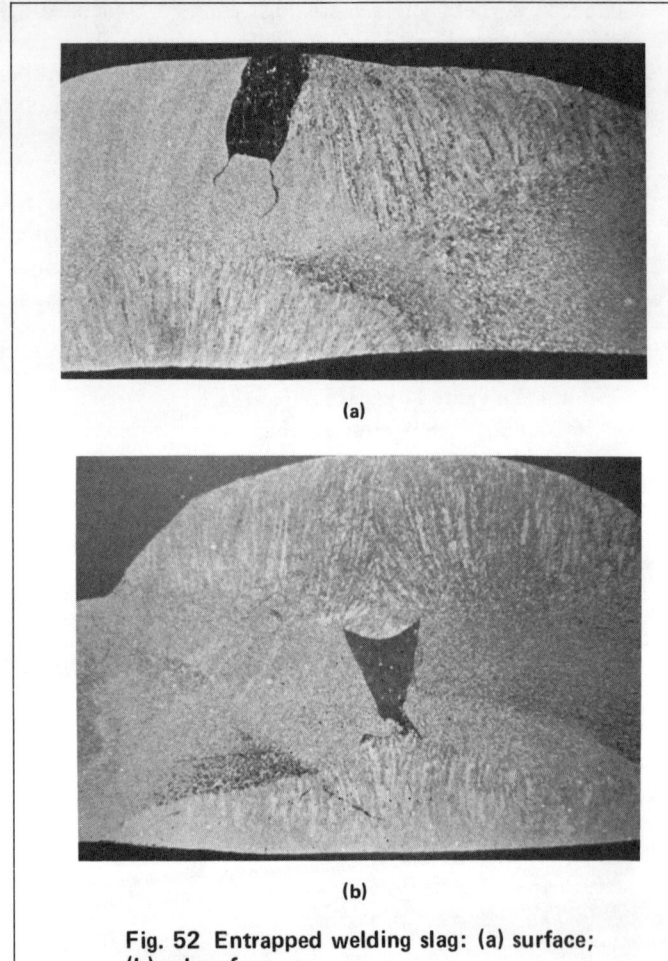

Fig. 52 Entrapped welding slag: (a) surface; (b) subsurface.

When observed, it can be explored quantitatively by grinding. It may be observed either at the root or entrapped near the top surface.

5. *Porosity*. Gases of various types and from various sources sometimes become entrapped in the molten weld metal. They form globular voids (Fig. 53). Gases result from cooling weld metal, changing solubility limits, chemical reactions in

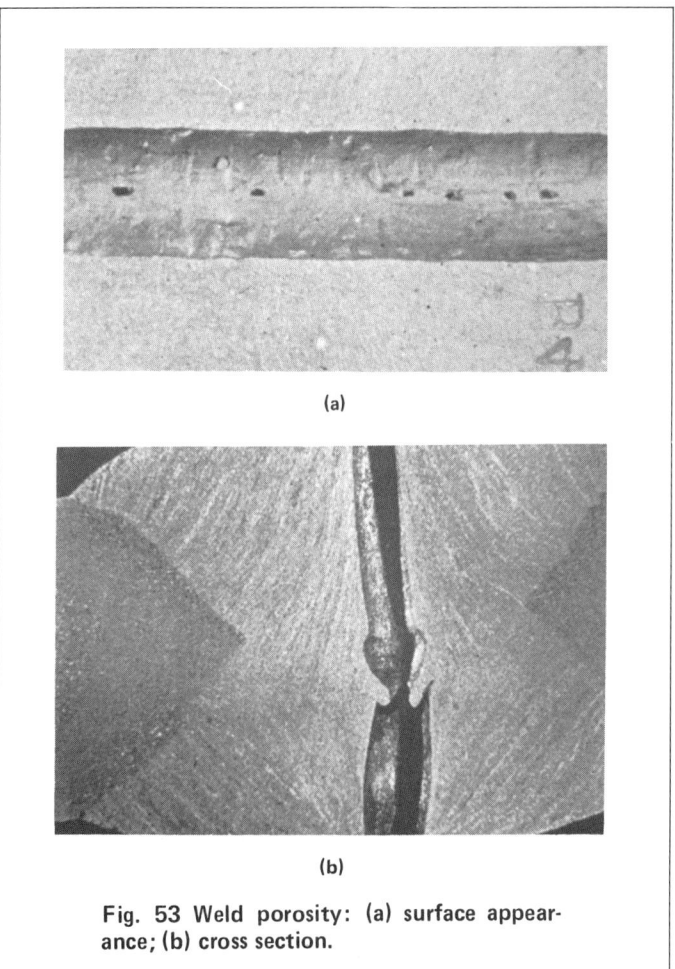

Fig. 53 Weld porosity: (a) surface appearance; (b) cross section.

the weld pool, and contamination. High temperatures, high amperage, and long arcs promote entrapped gas. Gas, like slag, is observable only if it surfaces.
6. *Weld cracking.* This is a term that is self-explanatory. There are, however, several forms of cracking:
   a. Hot tearing occurs down the centerline of the weld metal (Fig. 54a). The centerline is the last part to solidify; just after solidification it has little strength. If the parts

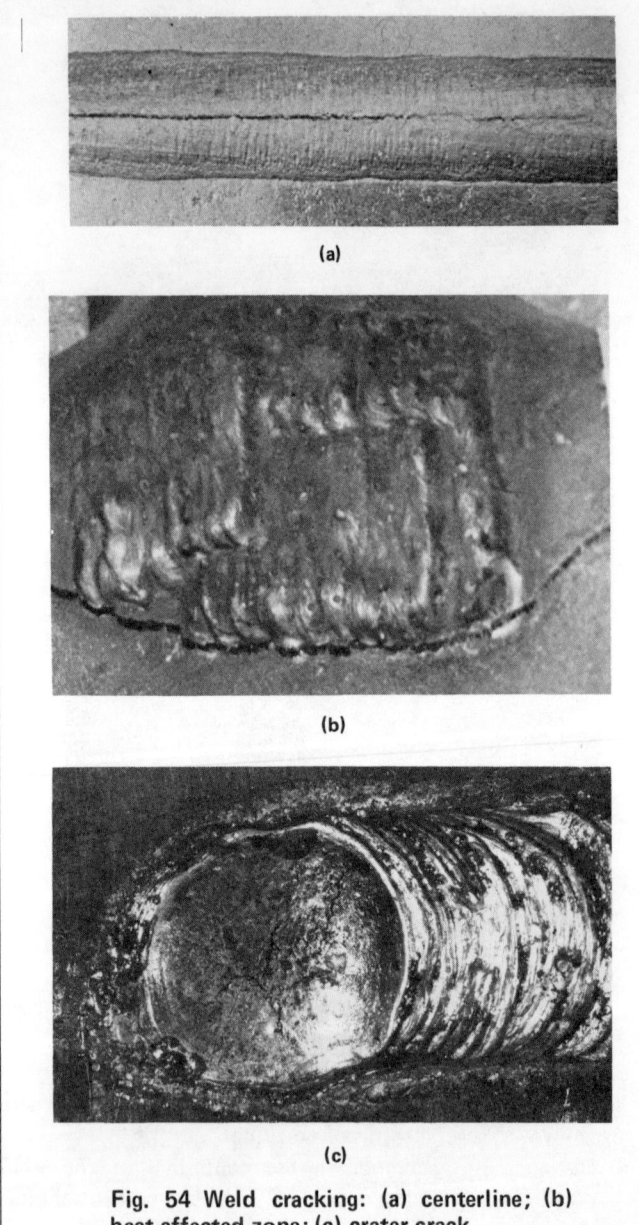

Fig. 54 Weld cracking: (a) centerline; (b) heat-affected zone; (c) crater crack.

being welded are restrained from normal cooling contraction, this weak weld centerline will crack. Such cracking is readily observable by visual examination.
   b. Heat-affected-zone cracking (Fig. 54b) occurs immediately adjacent to the weld metal. The heat-affected zone is that narrow band of base metal adjacent to the weld metal where the material is heated above the critical temperature. When it cools, a very hard, brittle phase occurs if the hardenability is high. This is particularly true of higher carbon steels. Such cracking generally surfaces and is observable. These cracks are often tight and require careful visual examination, with some magnification.
   c. Underbead cracking is usually caused by hydrogen. This type of cracking is visible less often, as it forms subsurface between beads and between weld metal and base metal. Such cracking leaves smooth fractures, which follow the natural contour of the weld bead.
   d. Crater cracks (Fig. 54c) occur in the sunken area at the end of a weld bead. If this sunken area, or crater, is not filled before the pass is completed, cooling of the outer edges produces stress cracking at the center of the sunken area.

   There are many other types of cracks which occur for various reasons. Lamellar tearing may occur due to lack of homogeneity of the material being welded (banding and nonmetallics). It is more common in heavy sections and exhibits a woody fracture appearance. Slag, gas, inclusions, and laminations are all capable of producing cracks. These can usually be readily related to their source. Some materials have low ductility, as do some weld deposits. These crack due to inherent brittleness. They may only be welded with great care, or in some cases, are not weldable. Locked-up internal stresses can cause cracking if stress relieving is not done quickly and properly, or not done at all. A good knowledge is helpful in ascertaining the likelihood and cause of such cracking.
7. *Arc strikes, arc drag, and weld splatter*. Although these are three different defects, the end result is similar. An arc strike (Fig. 55a) occurs when the welding rod contacts the metal being welded in an area away from the weld groove. It gen-

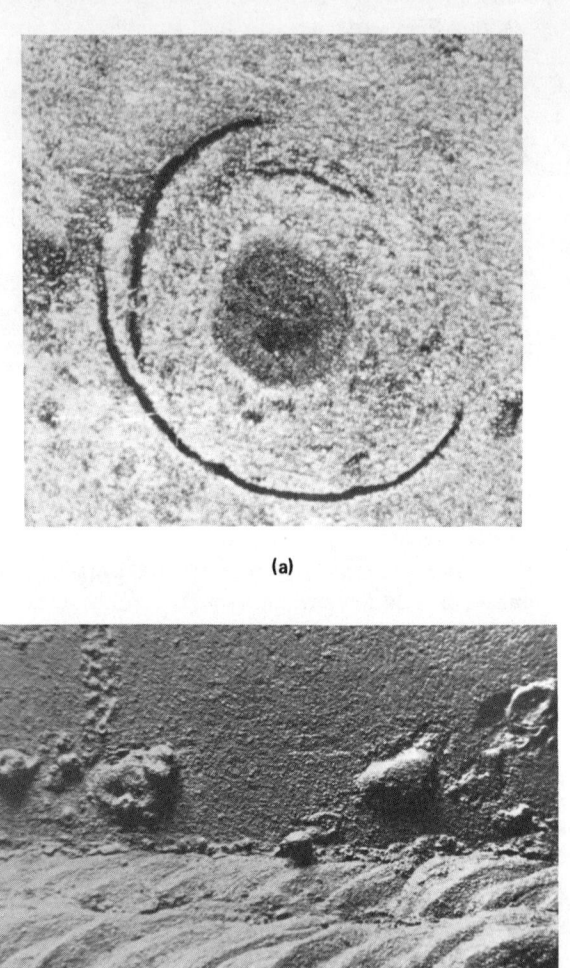

Fig. 55 (a) Arc strike with stress corrosion cracks surrounding it; (b) arc drag and weld splatter.

erates high heat in a small area for a short time. The adjacent cold metal cools this area rapidly, producing a hard, highly stressed localized zone. This zone is susceptible to cracking and corrosion. An arc drag (Fig. 55b) is an elongated arc strike producing the same effect. Weld splatter has much the same effect as an arc strike, but to a lesser degree.

8. *Excess penetration (burn through).* This results from high heat input, causing abnormal base metal melting. This causes excess metal build-up, wide heat-affected zones, and possible burned metal (Fig. 56).

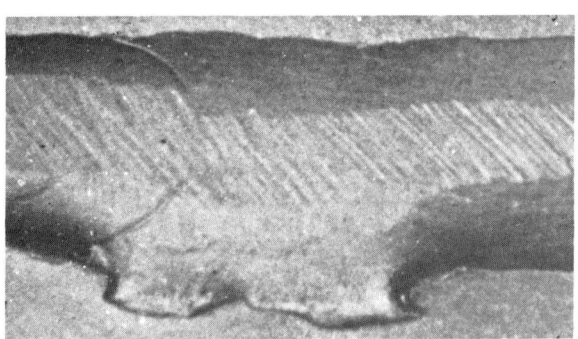

Fig. 56 Excess burn through at the root of a weld.

9. *Hard heat-affected zones.* The narrow zone adjacent to the weld metal is heated high enough during welding to harden on cooling. The hardness varies with type of material. Heat treatment after welding will soften this zone. If left in the hard condition, it is crack sensitive and leaves a ridge when machined (Fig. 57a).
10. *Entrapped foreign material in weld metal.* This can take many forms. The most common is when a section of electrode is put in a weld groove and covered with molten weld metal. This produces a nonfused area and often leads to premature failures (Fig. 57b).

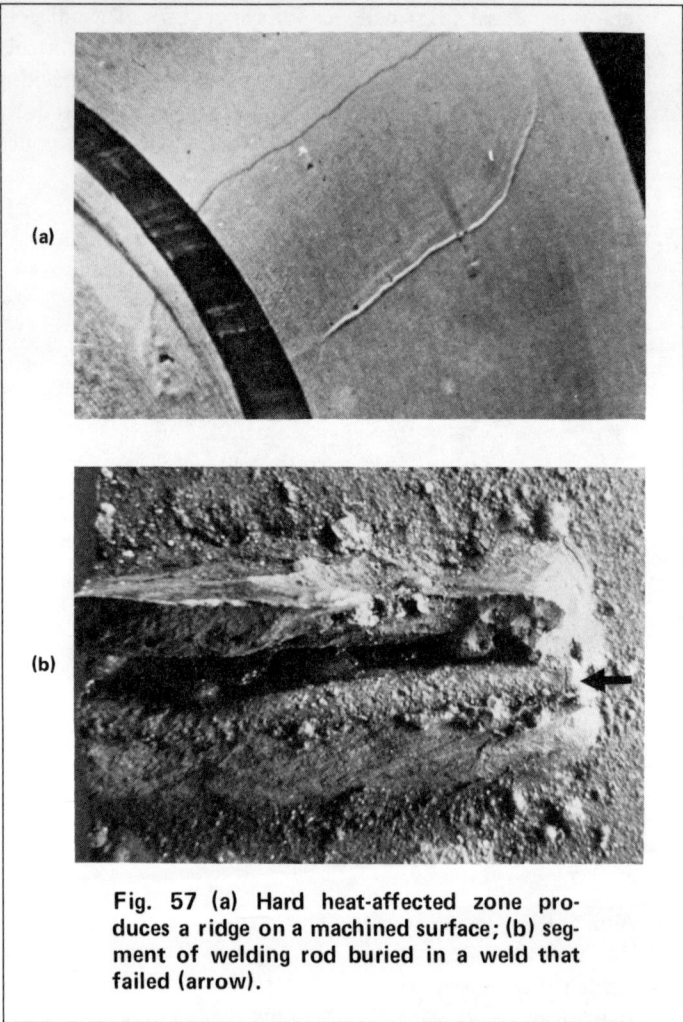

Fig. 57 (a) Hard heat-affected zone produces a ridge on a machined surface; (b) segment of welding rod buried in a weld that failed (arrow).

11. *Distortion.* When parts must maintain certain dimensional tolerances, distortion and warpage are considered defectiveness in the material. This problem can be avoided in two ways. The materials should be as stress-free (stress-relieved) as possible prior to welding. Welding heat can relieve stresses in a localized area only, causing upset distortion if the parts to be welded are highly stressed (not stress-relieved). The

other method of preventing such distortion is sequence welding. This involves welding first in one area, then on the opposite side, and alternating—much like the tightening of lug bolts on a vehicle wheel. Preheating, maintenance of interpass temperatures, and proper stress relief after welding may also prevent distortion.

Electrical Resistance Welding

There are many types of electrical resistance welding (ERW). Of these, the four most common are flash welding, upset welding, seam welding, and spot welding (Fig. 58). In ERW welds, fusion is accomplished by electrical resistance and pressure. The electrical resistance supplies the heat and is the only source. There are no fluxing materials or filler metals.

Spot and seam welding are similar in that seam welding resembles a series of overlapping spot welds. Upset welding and flash welding are similar also. Upset welding involves placing two prepared surfaces in intimate contact under pressure and passing a current across the gap until the fusing temperature is reached. Flash welding involves contacting two surfaces, causing an electrical short circuit and instant melting. Pressure is applied, creating an upset. Metal oxides and slag are forced out, creating a flash that is subsequently removed.

For all resistance welds, except flash welds, surface preparation is important. The surfaces must be clean and free of paint, oil, grease, scale, etc. The surfaces must make good, uniform, well-aligned contact.

Spot welds should be round (or in some cases oval), smooth, and have a good uniform appearance. Some of the defects that may be observed and their causes are as follows:

1. *Indentations*. These are reduction in metal thickness due to electrode shape, improper control of force, and high heat generation. This causes reduced strength and poor appearance (Fig. 59a).
2. *Misshapen weld*. Irregular shape can be caused by electrode condition, alignment, poor fit of parts, relative movement of parts, and lack of cleanliness. This may reduce the strength (Fig. 59b).
3. *Surface fusing and/or electrode deposit*. This covers an assortment of problems, all of which are visually obvious. When

Fig. 58 Appearance of four types of electrical resistance welds: (a) spot weld; (b) seam weld; (c) flash weld; (d) upset weld.

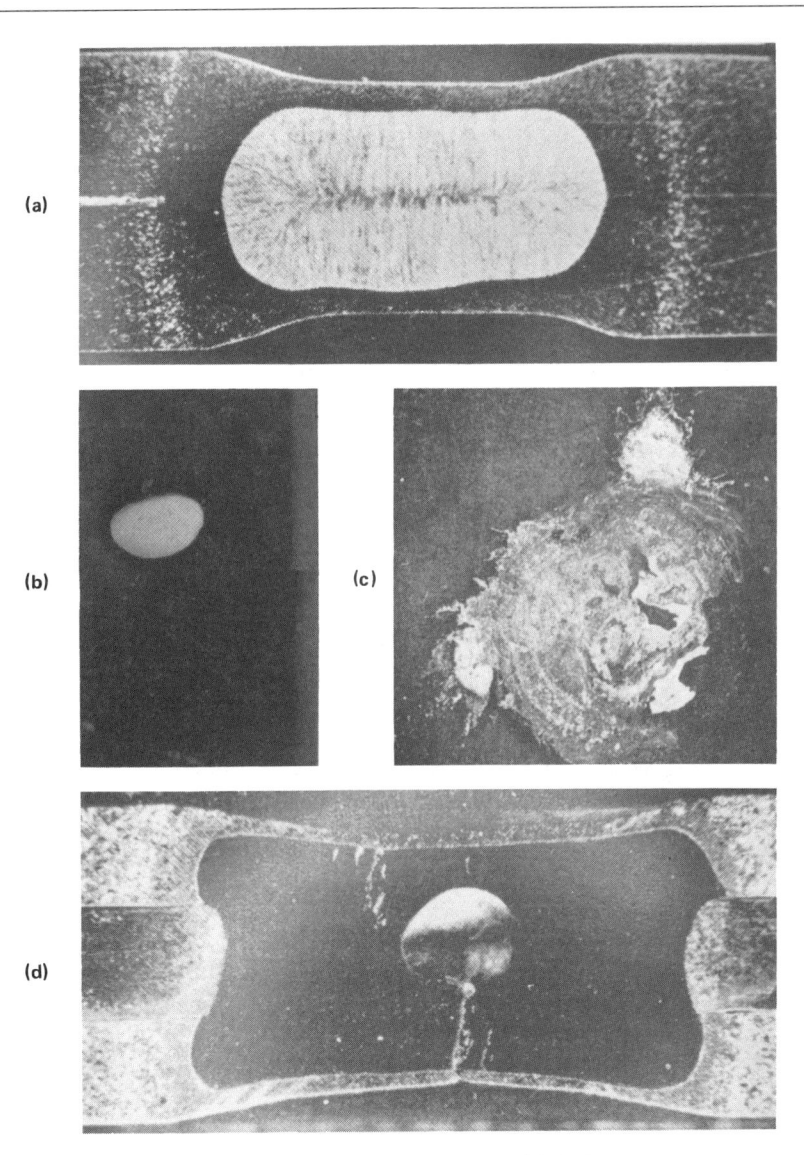

Fig. 59 Defects in spot welds: (a) indentations; (b) misshapen weld; (c) surface fusing and electrode deposit; (d) cracks and gas holes. (From Resistance Welding, published by American Welding Society)

part of the surface around the bead melts, the bead takes on a jagged shape. If electrode deposits occur, they appear as a foreign material. The electrode may also produce an indentation in the bead surface. These problems are caused by high welding current, poorly cleaned surfaces, low electrode force, and improper sequencing resistance. They result in lowered strength, smaller than normal welds, and shortened life (Fig. 59c).
4. *Cracks and gas holes.* Cracks and gas holes, mostly of the pinhole type, can occasionally be observed in the weld itself. These are caused by high heat, poor fit-up, and removal of electrode pressure too soon. They may promote corrosion, crack extension, or fatigue (Fig. 59d).

Another thing to consider is size of the bead. Formulas exist that relate the size of the visible weld bead to the fused area, based on the thickness of the material. Although proper size at the surface does not assure proper fused area, it is an important visual inspection aid.

Flash welds and upset welds are similar for the purpose of visual examination for defects. Flash welding produces more consistent quality than upset welding. In both cases, it is most desirable to weld parts having the same cross section. Upset welds have large upsets with gradual slopes; flash welds have smaller upsets with sharp slopes, with shorter temperature gradients. For both, the material should, if possible, be examined prior to welding. Flash and upset welding can cause the following problems (Fig. 60 and 61):

1. *Incomplete fusion.* Often visible as a cracklike line at the fusion point (Fig. 61b). May be caused by equipment malfunctions or contamination.
2. *Die burns.* Visible as localized arc type burns or melting (Fig. 61c). May result in cracking from corrosion or hard zones. Caused by improper contact between die and part.
3. *Entrapped oxides.* Not generally visible. Caused by insufficient upsetting. Flash appearance and slope angles may be cause to suspect (Fig. 61d).
4. *Craters and voids.* Visible upon surface examination and caused by improper equipment operation.
5. *Cracking.* This may or may not be visible. Cracks may develop

Fabrication / 97

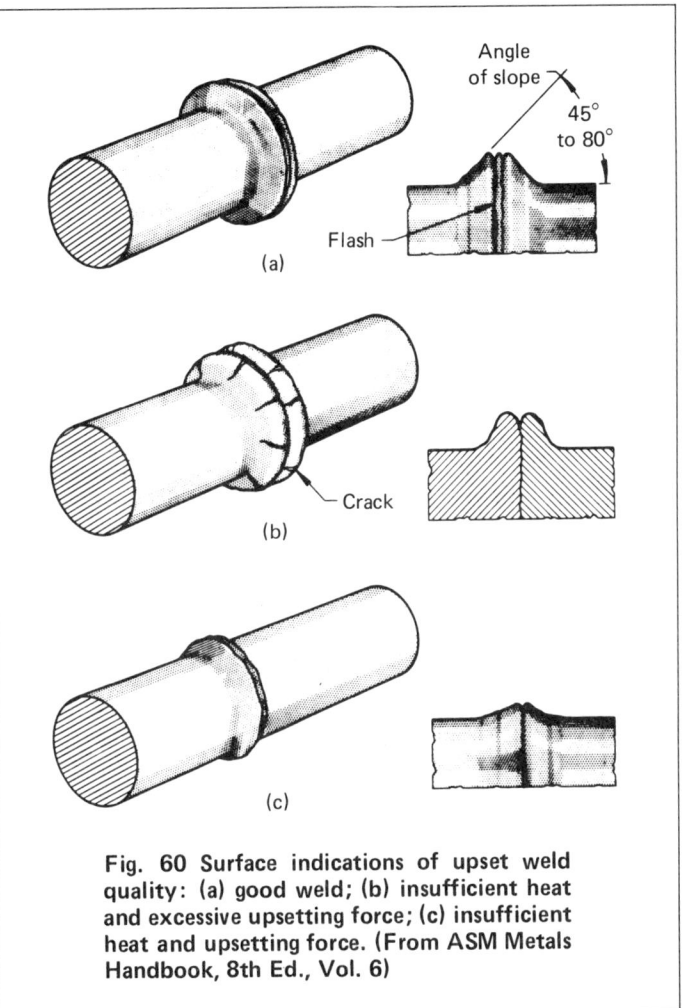

Fig. 60 Surface indications of upset weld quality: (a) good weld; (b) insufficient heat and excessive upsetting force; (c) insufficient heat and upsetting force. (From ASM Metals Handbook, 8th Ed., Vol. 6)

transverse or longitudinal to the weld. Cracking has different causes. Some materials do not weld well and will crack when heated. Some form hard, crack-sensitive zones after welding. If welding is attempted without sufficient heat, the upsetting operation may cause cracking. The slope and amount of flash may give a clue to cracking causes. Preheating and/or postheating may be necessary to produce a crack-free weld. This

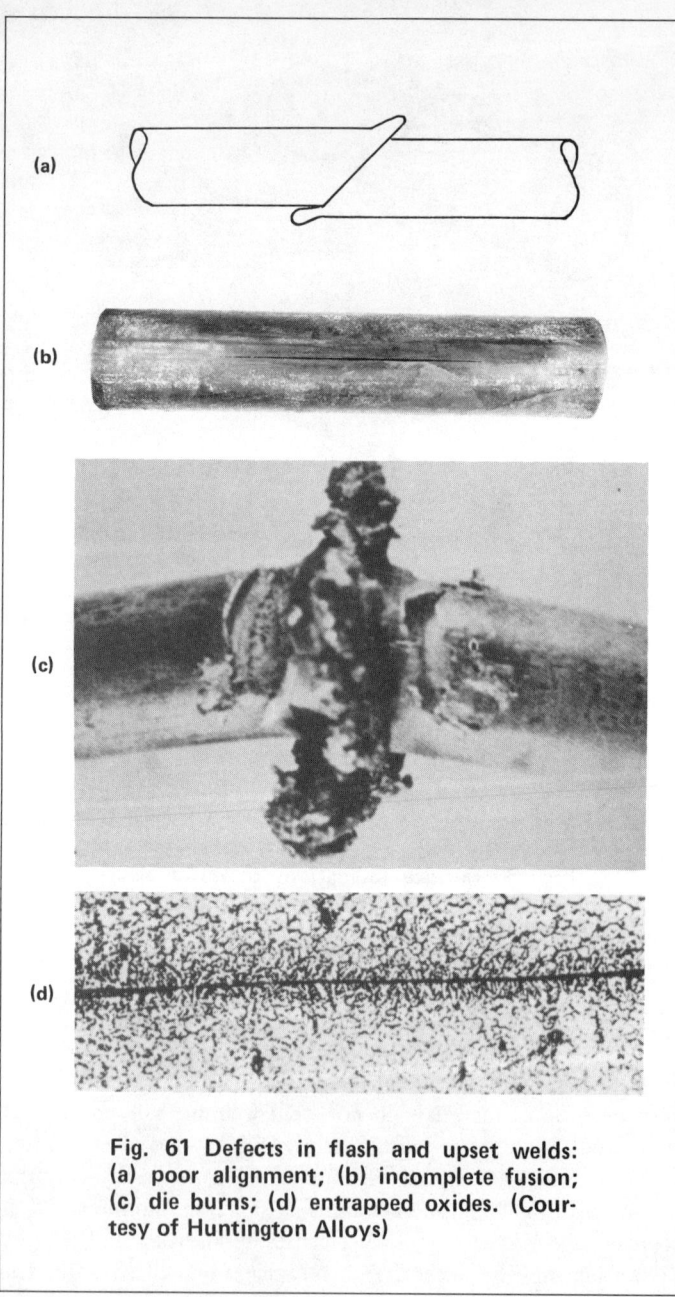

Fig. 61 Defects in flash and upset welds: (a) poor alignment; (b) incomplete fusion; (c) die burns; (d) entrapped oxides. (Courtesy of Huntington Alloys)

can be accomplished electrically as part of the overall process.
6. *Miscellaneous problems.* These are generally obvious, as are their causes and solutions. They relate to alignment of parts, uniformity of upset, amount and slope of flash, and upsetting. Mechanical adjustments are usually sufficient solutions. The heat pattern adjacent to the weld should be observed carefully for uniformity, color, and width. The surface adjacent to the weld should not be distorted.

Brazing

Brazing is a welding process that uses a filler metal that has a liquidus temperature (melting point) above 800 F, but below that of the metals being welded. The filler wets the surfaces of the welded metal and fills gaps by capillary action. It can be accomplished with a torch, in a furnace, or by any of a number of controlled heating methods.

Visual examination is probably the most common inspection method for brazed joints. Prior to brazing, the parts should be inspected for cleanliness and fit-up. Improper cleaning may result in voids caused by incomplete fusion. Too tight a fit may prevent flow-through of brazing metal, whereas an excessively large gap may not bridge over completely, leaving visible or invisible voids.

The most important after-brazing inspection is to ensure that the braze metal has completely filled all joints or gaps and has a sound, well-fused appearance. Brazed joints that show gas pockets that surface are probably not sound internally (Fig. 62a).

If excessive heat is applied, erosion of the base metal can occur (Fig. 62b). The flux used in accomplishing a bond may become entrapped in the joint. This is a particular problem where excessive oxidation occurs. Flux residue should be removed after brazing, as it is often corrosive.

Soldering

Soldering, while often thought of as a method for joining copper wires in electrical applications, is much more widely used. The following is a list of soldered materials:

        Copper   Stainless steel
        Copper alloys   Nickel

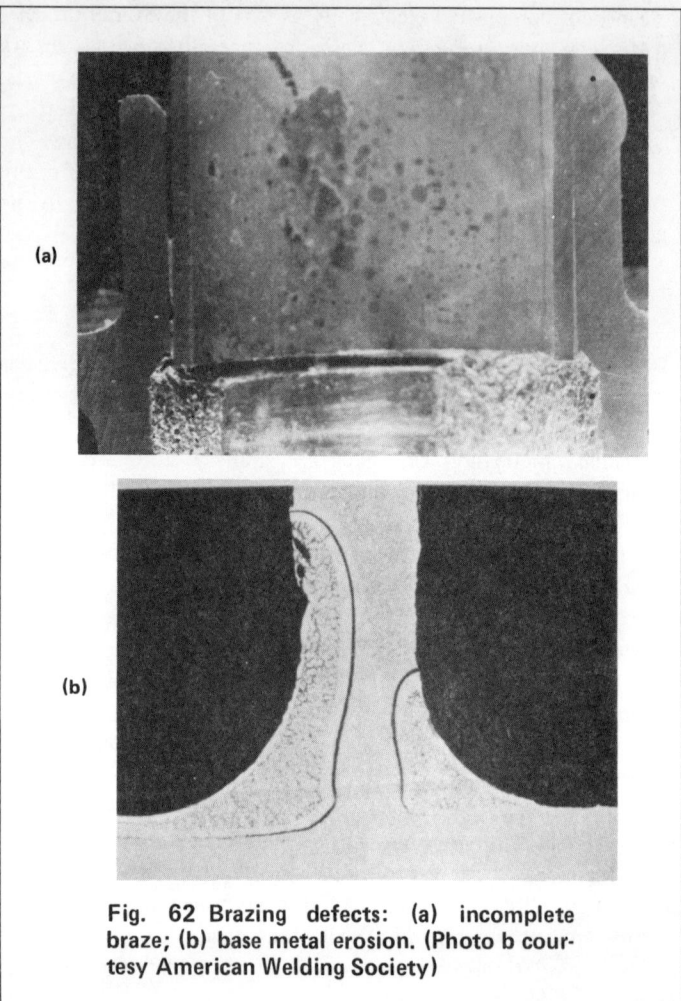

Fig. 62 Brazing defects: (a) incomplete braze; (b) base metal erosion. (Photo b courtesy American Welding Society)

Steel	High-nickel alloys
Coated steels	Lead alloys
Aluminum alloys	Magnesium alloys
Tin	Cast iron
Precious metals	Printed circuits

Visual examination is one of the best and probably the most widely used inspection method for soldered joints.

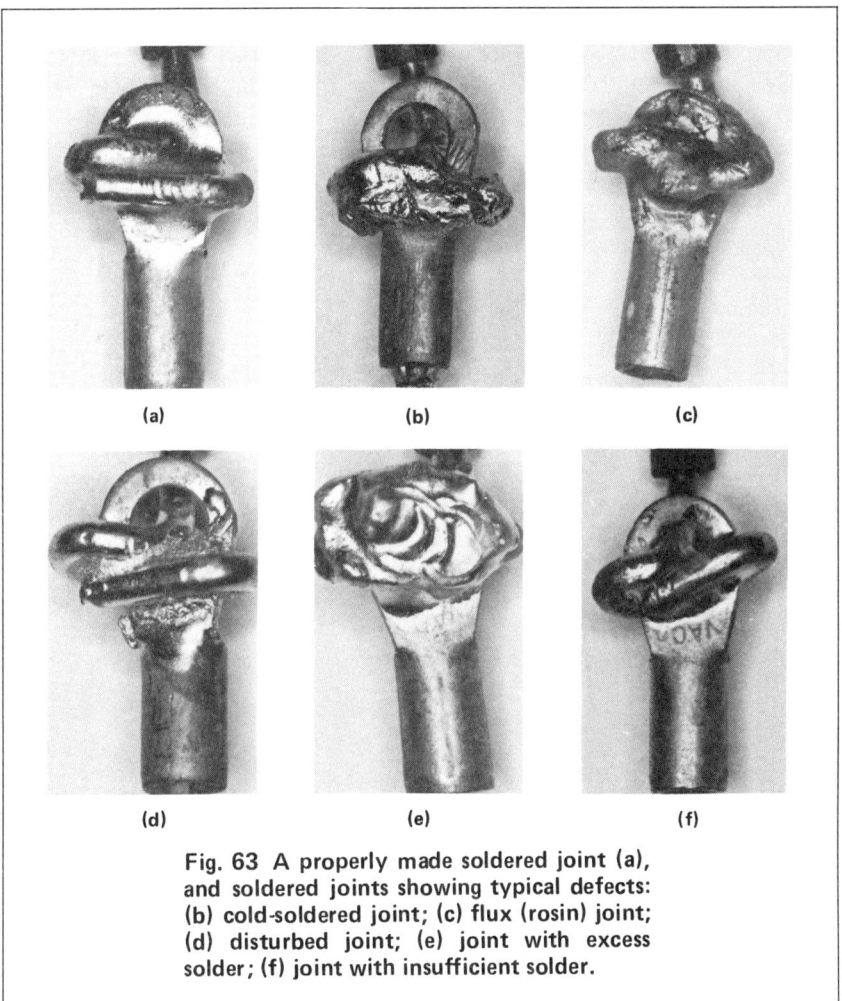

Fig. 63 A properly made soldered joint (a), and soldered joints showing typical defects: (b) cold-soldered joint; (c) flux (rosin) joint; (d) disturbed joint; (e) joint with excess solder; (f) joint with insufficient solder.

Typical soldering defects and their causes are:

1. *Cold-soldered joint.* Solder does not adhere to the metal being soldered. This is caused by improper heat application with the base metal not being heated sufficiently (Fig. 63b).
2. *Disturbed joint.* If movement occurs between the parts being soldered before solidification is complete, a faulty joint will result (Fig. 63d).

3. *Flux (rosin) joint.* If a flux layer becomes entrapped or forms at the interface between solder and parts being soldered, a faulty joint will result (Fig. 63c). This is caused by insufficient heat to dissolve and float out the flux.
4. *Excess and insufficient solder.* If too much solder is used, the quality of the joint is difficult to determine and the overall appearance is bad (Fig. 63e). If not enough solder is used, weak marginally attached joints result (Fig. 63f).

Visually, these soldering defects cause the following physical appearances:

1. *Cold-soldered joint.* Chalky crystallized appearance, not bright appearing with uniform flow characteristics, often exhibiting a high bead crown.
2. *Disturbed joint.* Shows evidence of localized deformation of the solder and localized, hairline lack of attachment.
3. *Flux (rosin) joint.* Residual flux can be observed at the interface and generally present on the surface.
4. *Excess or insufficient solder.* These defects are fairly obvious, particularly if a proper joint can be used for comparison. The excess solder forms an unattractive appearance; the lack of solder shows the joint to be only partly attached.

CHAPTER 5

Inspection of Manufactured Products

Justification of Inspection

Many manufacturers regard inspection as an overhead item. In itself, it produces no income; in fact, it is a major expense. But inspection produces satisfied customers and thus increased business. It lowers overhead since parts are rejected before the expenditure of expensive processing time. It lowers rejection rates by communicating to suppliers that such parts will not be accepted. This improves the overall quality of incoming materials. Inspection can help prevent product liability lawsuits, which today can be a matter of company survival.

With proper training and understanding, each manufacturing operator can become an inspector. To most operators, rejects are a problem. It is important that in rejecting parts they know something is being done about the rejection rate. A typical comment by a machinist is "I told them for a long time that certain batches of these forgings had hard spots and were tearing up my tools. I know these tools are expensive, but no one seemed to care or do anything about it, so I just quit saying anything." Some inspection training for operators can reduce inspection overhead. A book such as this could be used as a training aid.

Locating defective components is only part of the job of inspection. Tracing the cause of the rejection results in a general upgrading of quality. A good example of what a strong inspection program can do is demonstrated by the following actual incident. A manufacturer purchased similar parts from a number of suppliers. The defect rate was high, entailing expensive repair and rejection. The suppliers were called in to discuss the problem. An agreement was made in which a premium of 25% was paid for additional inspection, but any defective part not rejected by the supplier and discovered by the manufacturer would be repaired by the manufacturer, with all costs backcharged to the supplier. While the 25% premium sounded reasonable to the suppliers, they soon discovered that backcharges exceeded this premium. Another meeting sent inspectors from the user to the supplier to train his inspectors to inspect parts properly and to reject unacceptable materials. This resulted in a dramatic reduction in repair backcharges. Soon the purchaser's name was associated with stringent quality at the supplier's plant. Marginal batches of products were diverted to customers who were not so particular. The 25% premium was then renegotiated to 5% with no backcharge provision, resulting in very high as-received product quality for many years. Inspection is a matter of understanding and communication among all parties involved.

Specifications

Quality originates with specifications. A product is only as good as it is specified to be. It cannot be effectively inspected more closely than has been specified. The acceptable and unacceptable limits must be defined. In the absence of specifications, the courts have generally found that a product should be manufactured so that it will perform its function properly, for a reasonable period of time, when used in an expected manner. This is probably a good beginning outline for writing a specification.

A specification should be practical. It cannot be more stringent than the manufacturing capabilities can produce. On the other hand, a good specification has to be enforced. There are many specification-issuing bodies, including ASTM (American Society for Testing and Materials), ASME (American Society of Mechanical Engineers), and API (American Petroleum Institute). Their specifications can be adopted and used as is, or they can be modified. Most of them have provisions for modification, including the general statement "or by agreement between the manufacturer and purchaser."

For the best final product, it is wise to write a stringent, but practical, specification. Specifications should be discussed with the manufacturer. A premium will have to be paid for those special qualities considered essential. Documentation of compliance should be demanded from the manufacturer for each batch produced. When the product is received, it should be inspected by the authorizor of the specifications. *Do not expect the manufacturer to do your inspection for you.*

Inspection at the Supplier's Facilities

With a new product or supplier, it is well to send an inspector into the supplier's plant where possible. This allows both parties to communicate their expectations. The purchaser knows what the supplier is capable of producing, and the supplier knows what sort of inspection the parts will receive upon receipt. Good in-plant inspection—done initially and at least periodically—can develop a respect for your product and often foster preferential treatment. Visual examination is the principal inspection method used at a supplier's facility. This can also be an opportunity to train the supplier's inspector to look for the same things you do.

Incoming Materials Inspection

All parts should be inspected upon receipt. Immediate rejection of parts is important for two reasons. Quick rejection helps solve the problem before more rejectable parts are produced. The manufacturer is more likely to accept a quick rejection. Most incoming inspection takes the form of visual and dimensional examination. All the tools and methods discussed in Chapters 1 and 2 should be considered.

In-Process Inspection

Inspection should not be limited to incoming products. It should include inspection during processing and of the final product.

In-process inspection may occur at several logical manufacturing stages. Inspection after rough machining reveals defects covered up by a surface skin and by rust and corrosion. Cracks, porosity, and weld repairs are often revealed.

If heat treatment is a process step, inspection after heat treating may reveal defects which were either caused by the heat treatment or accentuated by the heat treatment. Heat treating can produce cracking that can be very tight (fine-lined). Careful examination with some magnification may be necessary. Heat treatment can produce drastic dimensional changes as the result of relieving locked-up stresses.

Another good place for in-process inspection is after welding. The many problems that occur during welding and other fabrication methods are outlined in Chapter 4.

While most problems are observable upon inspection of the finished product, much expensive processing can be avoided by early rejection.

Finished Materials Inspection

Inspection of the final product may be the most revealing because the surface finish is more conducive to exposing defects. One hundred percent inspection is desirable, but not often practical.

There are many books and articles on quality control that can be used to develop inspection procedures for determining the number of parts of a batch to be examined to reasonably ensure satisfactory production. Whenever a defective part is found, the number of parts inspected should be temporarily increased for assurance that it is not a repetitive defect.

Field Inspection

Field inspection is similar to all other inspection procedures previously described, except that it must often be done under the most adverse conditions. The temperature may be too hot or too cold. It may be necessary to work in rain or in damp areas. Lighting is often poor. Electricity may or may not be available. Space is sometimes limited, as is access.

Inspection of Manufactured Products / 107

A field inspection kit is recommended if much of this work is to be done. The kit may include items discussed in Chapter 2. Of these, a good battery-powered flashlight, clipboard, magnifiers, and camera are basic necessities. Coveralls are also recommended for clothing protection. (For a discussion of a typical field inspection kit, including itemization of all equipment, see Chapter 6.)

It is well to inquire before inspection about the conditions—lighting, access, available electricity, and local assistance—under which the inspection will be made.

CHAPTER 6

Failure Analysis

Failure analysis is becoming increasingly important, for various reasons. Some failures are quite costly. Not only is there a loss of valuable equipment, but product output interruptions may occur. In modern high-volume production facilities, downtime can cost tens of thousands of dollars per hour.

Not long ago, the cost of materials used in plant equipment was quite low. If a part failed, it was scrapped without a second thought. With the high prices of both materials and labor today, the cost of determining the cause of failure can well be justified. Information so developed can then be used to prevent future expensive failures. It can also be used to fix fault and perhaps get supplier compensation for the faulty product.

An even greater emphasis on failure analysis has occurred due to changes in the laws involving product liability. The newspapers are full of stories about the large awards to victims of products that have failed in use. Although these are not confined to metal products, a great many do involve metal failures.

Background Information

Analyzing metal failures is like playing detective with inanimate objects. The first step is gathering data, sometimes known as "background information." This includes any available information relating to what occurred leading up to, and especially just prior to, the failure.

Information falls into two categories. Hearsay information is a scenario of the events as described by eyewitnesses. Any such information furnished should be weighed carefully, since it often involves bias. It has been shown that various witnesses to the same event tell widely differing stories as to what they observed. Witnesses tend to embellish the event. They also tend to relate what they assume you want to hear. They think in terms of protecting themselves and their fellow workers from blame.

The second, and far better, way to gather information, is the acquisition of such detailed factual data as *date* and *time* of the failure, *weather conditions*, *written reports*, *charts*, and *photographs* made at the scene. The following is a list of some of the operating data that may be gathered:

>Material specifications
>Design operating conditions
>Length of service
>Temperatures and pressures
>Static and dynamic loading
>Corrosive and erosive conditions
>Cyclic loading
>Testing, inspection, and maintenance
>Any unusual occurrence.

Tools of the Trade

If failure analysis is done with any degree of frequency, it is helpful to prepare a field inspection kit. It may take the form of a tool box or a fishing box. The following is a list of suggested items to be included in such a kit:

Magnifying glass	6-in. slide calipers
Tape measure	Felt-tip markers
Inspection mirrors	Modeling clay
Cotton swabs	Knife
Solvent	Mill file
Small hammer	Tweezers

Small rule*	Screwdriver
Masking tape	Scissors
Labels and tags	Magnet
Sample envelopes	Scribe
Sample bottle	Brush
Replicating tape	Abrasive paper
Needlenose pliers	Stethoscope
Small metal-cutting saw	Safety glasses
Flashlight (rechargeable)	Colored wax markers

*Coils of ruled material with self-adhesive backing are available. This can be cut to length and attached to parts being photographed to show size.

A stereoscopic microscope, while it will not fit into a field kit, will give the capability of higher-magnification examination.

A second kit should contain camera equipment. The 35mm single lens reflex camera is the most popular, and has many optional features. There are other satisfactory cameras, including the view cameras that make superior photos, but are large and bulky. In addition to fixed-image cameras, consideration should be given to both motion picture cameras and video cameras. Most video cameras have instant playback capability which assures proper coverage. Both motion picture and video cameras are available with sound tracks for recording at the scene or narrating later.

The ideal camera kit is padded to protect the equipment from travel and the usual field abuse. This kit should contain camera equipment suitable for making records ranging from extreme close-ups to telephoto capabilities. Interchangeable lenses provide this versatility. Adequate supplies of various films should be included. Films to consider are black-and-white, color print and color slides. Both fast and slow films are important, as lighting varies, things move, and enlargements may be required. A flash attachment is a necessity. Antiglare and lens protective covers are desirable, as are an assortment of filters. A sensitive, separate exposure meter is helpful because of the great contrasts that may exist between general lighting and the usually darker parts being photographed.

Examination is not complete without good, accurate record keeping. Record keeping in field examination can be a problem. The area may be dirty, hot or cold, outside, windy, and rainy—conditions not conducive to taking good notes. A clipboard with a cover is desirable. Graph paper may be useful in making drawings to scale or in perspective. A pocket recorder is a convenient way of taking field data. It either complements or eliminates much note taking.

Preparations for Sample Removal

Since samples will probably be removed for laboratory examination, it is well to make arrangements ahead of time for such removal. There should be an understanding between all parties on the extent of sample removal.

Inquire about the on-site availability of removal equipment and assistance in such removal. Material handling equipment and acetylene torches may be necessary. If field power is nearby, an extension cord (100 ft or more) is a handy item (this can also include a light).

For smaller samples, a power saber saw, a hand hacksaw, a motor tool with small abrasive cutoff discs, hole cutters, and trepanning tools are useful. For larger samples, torch cutting may be necessary. A squeeze bottle full of water can be used to keep the sample cool during abrasive cutting.

In the event the part cannot be cut, the replicating tape in the field kit can be used to make very accurate replicas of such areas as the fracture origin.

Although it may be out of the realm of visual examination defined as nondestructive, field equipment is available for polishing, etching, and microscopic examination at the site.

On-Site Inspection

It is most desirable to examine the failure and site as soon as possible after the occurrence. Unfortunately, it is more often days, weeks, months, or even years later before such an examination can be made. With delayed examinations, it is very important to determine what has happened to the site and the parts since the failure. Where have the parts been stored? Under what conditions? A broken metal part left exposed to the elements may be drastically altered and clues to the cause of failure may be erased.

Start with a general examination of the scene. Observe the surroundings. Is there possible interaction with the surroundings that could have had an effect on the failure? The failure may have occurred in a low area where a combustible mixture could accumulate. There may be a nearby source of ignition. There may be nearby construction that produced out-of-the-ordinary conditions. Be observant, ask questions, examine pressure and temperature gages. Check on the weather conditions at the time of the accident. Try to ascertain if any abnormal conditions occurred. Examine other similar units or parts if they are available. Observe any interacting parts or units to see if they show signs of playing a part.

Examination Procedure

Failure analysis begins with visual examination. This is not only the first, but also the most important, step. Visually examine the parts on all surfaces. Start with the naked eye, then increase magnification using a magnifying glass or visor. Progress eventually to higher magnification: first to the stereoscopic microscope, then, if necessary, to the electron microscope.

Try to account for all markings, deformations, abrasions, and discolorations. Observe and record any identification markings. The general manufacturer's nameplate and nameplates of the manufacturers of the components may be present. These may also be stenciled with paint or stamped or cast on. They may take the form of a symbol such as a trademark. There may also be heat numbers that identify a particular batch of metal. Look for operating instructions, warnings, and limitations.

Look for evidence of abuse. Try to decide if abuse occurred before the failure or after (see Fig. 6, in Chapter 1). Put yourself in the position of the part and visualize how it might have acted in coming apart. Did it deform, indicating overloading, or did it break in a brittle manner? If deformation occurred, what type of loading does it indicate? If brittleness is observed, is the material likely to be inherently brittle, or does this denote an abnormal mechanism of failure?

A small file will reveal something about the hardness of the material. If abnormal brittleness is observed, try to determine its cause. Is there a pronounced stress concentration that can produce severe triaxial loading conditions? Is there evidence of corrosion, such as pitting, scaling, or discoloration? Could high or low temperatures have played a part? High temperatures may be revealed by scaling or temper colors (see Fig. 7, on insert between pages 54 and 55, and Fig. 8, in Chapter 1). If corrosion is observed, is it general or localized? Is it uniform or selective (pitting)? Is it in an area of contact with other materials? Does it have special characteristics indicative of high-velocity flow (erosion, cavitation)? Does it leave an unusual appearing corrosion product? It may be a good idea at that time to remove scrapings from specific areas to preserve and label for future tests. When outside and inside areas exist, as in tubing, do not mix scrapings from the two areas. Try not to mix grease, paint, coating material, and mill scale with the corrosion product sample.

Fracture Examination on Site

The fracture examination is more meaningful after the general examination. This examination phase is best accomplished in the laboratory, but certain things can and should be done in the field. An effort should be made to locate the origin or origins. If the fracture is large, these are the areas that should be cut out and removed to the laboratory.

There is an interesting mapping technique that can be used on large parts that have become severely distorted in the process of failure. This technique involves mapping of the fractured parts and superimposing these maps on a new part of the same size to show fracture paths. An example would be a large-diameter pipeline or cylinder that ruptured, producing various severely distorted pieces. Large sheets of paper can be taped to the distorted parts and cut to conform to the shape and fracture path. Other characteristics, such as welds and indentations, can be marked on the paper. The paper maps can then be pieced together on a cylinder of similar size and shape to the one that failed. This composite will show all fracture paths, welds, and marks relative to the original shape. It can be sketched, recorded photographically, and transferred to a small cardboard cylinder as a model (Fig. 64).

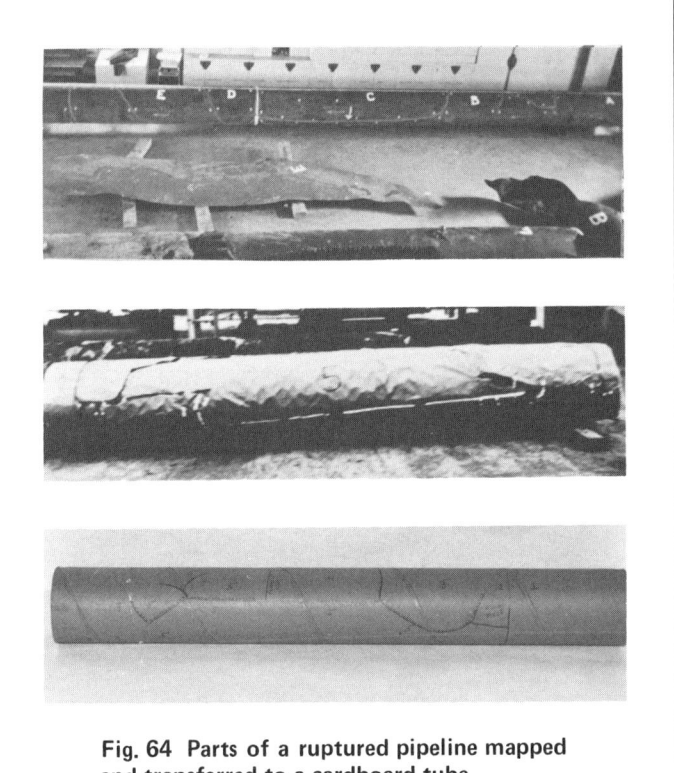

Fig. 64 Parts of a ruptured pipeline mapped and transferred to a cardboard tube.

The fracture should be characterized. Is it a single-phase or multiphase fracture? That is, are the characteristics of the fracture all of one type, or does the fracture change characteristics as it propagates? Location of the origin point or points of fractures will reveal much about why the fracture started and the direction of initiating forces. Often a fracture initiates by one mechanism and propagates by another. This initiating area may be large enough to be obvious or exceedingly small. It is more important to know the mechanism of the origination than that of propagation. Some fractures produce chevron- or arrowhead-type markings indicating the point or points of origin (Fig. 65).

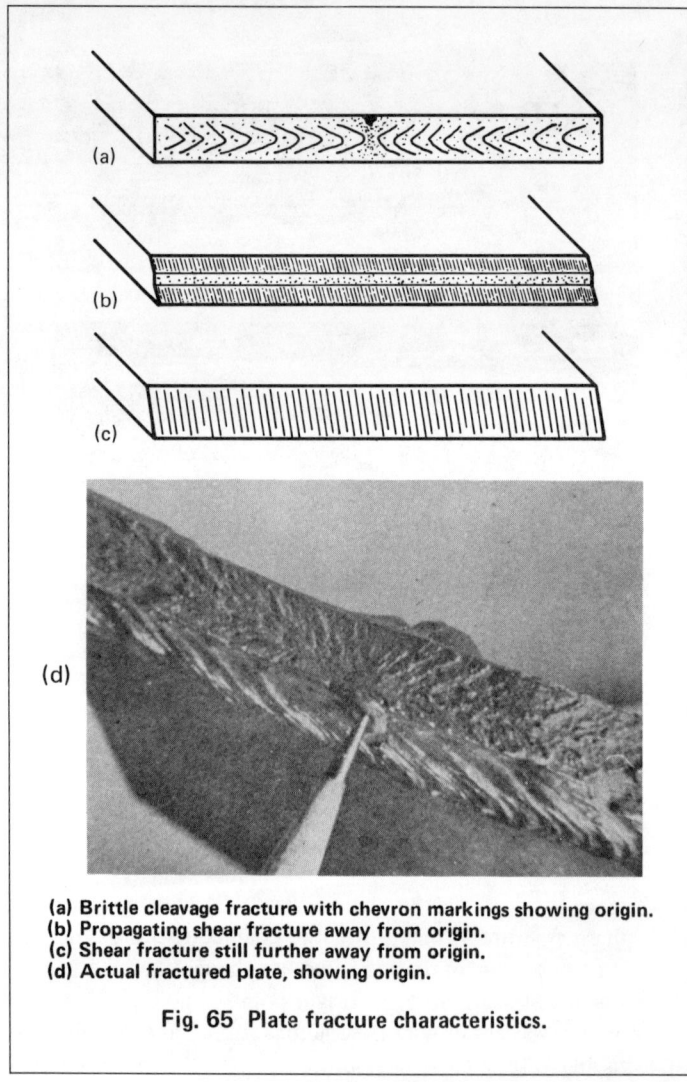

(a) Brittle cleavage fracture with chevron markings showing origin.
(b) Propagating shear fracture away from origin.
(c) Shear fracture still further away from origin.
(d) Actual fractured plate, showing origin.

Fig. 65 Plate fracture characteristics.

Measurements, Sketches, and Notes

Measurements fall into two categories: measurements made of the scene (general) and measurements made of broken parts (specific). Measurements, sketches, and notes all really complement one another. In addition to the sketches, extensive notes should be taken during visual examination. Do not depend on memory.

Start by sketching the overall area. You do not have to be an artist. The importance of sketching is that it forces you to see things you would otherwise miss. For example, take any object, look at it carefully, then sketch it in detail, noticing all the new things you see. Notes and dimensions should be added to the sketch. It is better to collect too much data than too little.

After completion of sketches, notes, and dimensions of the site area, repeat the same procedure on the broken part or parts. Try to ascertain the type, direction, and intensity of loading that caused the failure. Note if distortion occurred prior to failure. After taking notes, and making sketches and measurements, make photographic records. At this point, you will know much better which photos to take rather than shooting at random.

As soon as possible, in a place conducive to writing and sketching, redraw rough sketches and rewrite notes. This will tell you if you really have "seen." This also allows another person to work from your notes. Rewriting and sketching should be done while the information is fresh in your mind.

Sample Selection

Field examination usually ends with removal of samples to the laboratory for further, more detailed study. If small, the entire fractured part may be removed. Remove as much of the fracture as can be handled, or is permitted to be removed. The removed parts should include all origination and potential fracture origination areas. Removal should be such that the material is not altered.

When abrasive cutting or torch cutting is utilized for sample removal, the cuts should be made sufficiently far back so that any heat generated by cutting does not affect the areas to be examined.

The fractures and other areas should be protected during removal to avoid contamination by the cutting procedure. The importance of protecting the fracture as soon as possible—after it occurs, during removal, and in transit—cannot be overemphasized (Fig. 66). A heavy layer of plastic taped in place works very well. It is unwise to protect fractures with oil or grease or any other tightly adhering material, because this will contaminate surfaces, preventing corrosion-product analysis, and may attack the metal.

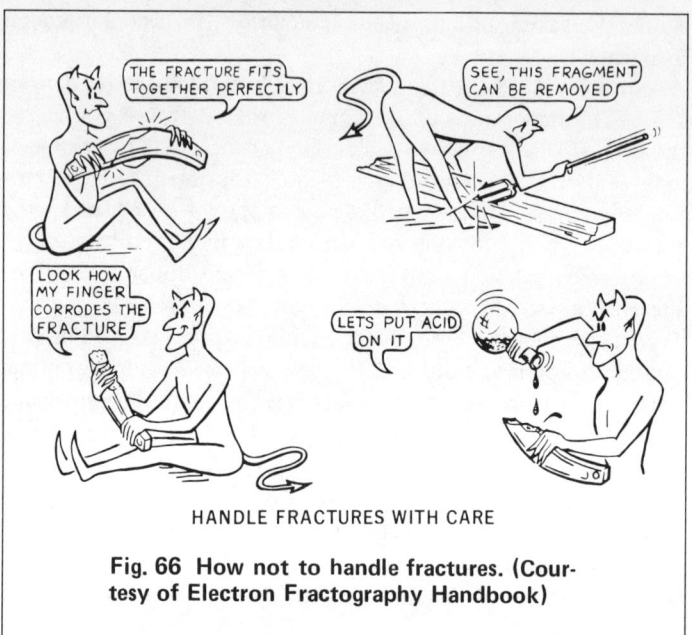

Fig. 66 How not to handle fractures. (Courtesy of Electron Fractography Handbook)

Never clean surfaces by such abrasive methods as wire brushing. Do not fit fractures together; this abrades the surfaces and hampers future fractographic examination, particularly at high magnifications. For the same reasons, never acid clean surfaces. It is best to confine field cleaning to a soft bristle brush. Further cleaning can be accomplished in the laboratory.

A number of sealable plastic bags of various sizes are good for protecting smaller parts. All parts should be carefully labeled. It is helpful to label the part, mark the area where cutting is to take place, and photograph prior to cutting. This will identify where each specimen came from.

Sometimes a part cannot be removed. It may be in an area of difficult access or it may be necessary to repair it quickly to resume service. In this case, more extensive cleaning and examination are required. Replication techniques produce very accurate reproductions of fractures. Such kits are available from laboratory supply companies. They involve softening of plastic tape in a solvent and carefully pressing it on the surface area to be examined. When it dries it will pop loose readily, giving a nondestructive, exact replica of any surface. If the sample is not clean, the technique may be repeated until all surface contamination is stripped loose. These contaminated tapes should be saved for analysis of the product pulled loose.

Another nondestructive technique is field metallography. This involves portable polishing equipment and microscopes. With this technique, the microstructure can be observed and photographed at the site, or replicas can be made of the polished and etched surface and observed in the laboratory.

Visual Fracture Examination

Laboratory examination of fractures is similar to general laboratory examination in that it utilizes the various magnifying devices. It also uses the micro-analytical techniques such as the microprobe. Results of these procedures are recorded photographically.

The first breakdown in fracture identification is the differentiation between ductile and brittle (Fig. 67). To do this, we must know something about the inherent ductility of the material from which the part is made. For example, most gray iron and very hard steels are inherently brittle and will produce a brittle fracture no matter how they are loaded. This is not true of other materials. Steels, unless they are at very high hardness levels, will fracture in a ductile manner. That is, they will fracture in such a manner if they are exposed to normal overloading.

Ductile failures or fractures are associated with loading in excess of the strength of the material (normal overload). This is indicative of one of three causes:

1. Accidental or purposeful overloading (the most common cause).
2. Improper design such that normal expected loading can produce failure.
3. Inadvertent use of materials of lower strength than called for in the design.

The common characteristics of all ductile failures are distortion, deformation, flow, and necking down prior to fracture. Examination can reveal much about the loading direction in ductile failure. Tension failures produce 45° shear planes and cup-cone fractures. Shearing failures produce the appearance of a knife cut-

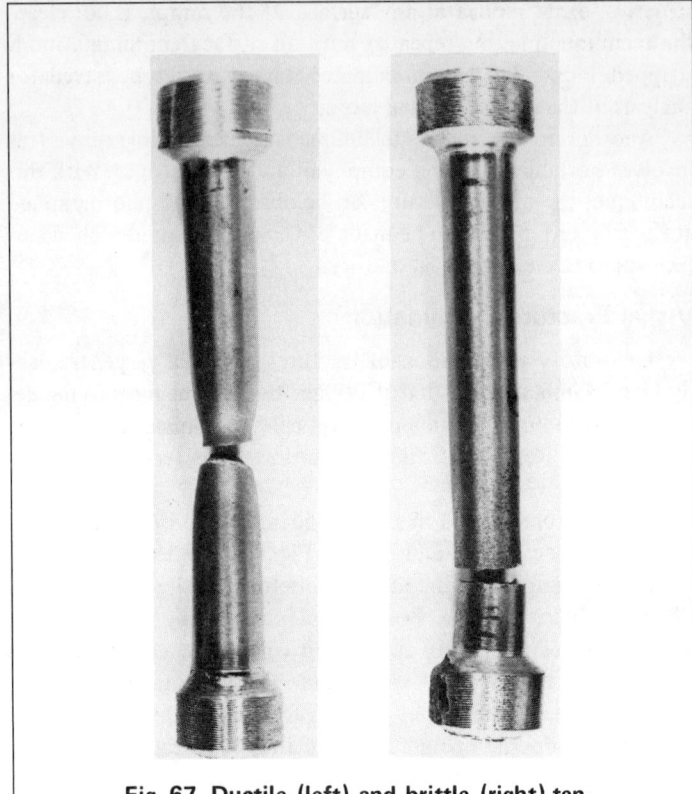

Fig. 67 Ductile (left) and brittle (right) tension-test bars.

ting soft butter. Torsion failures show an obvious twisting distortion. Compression failures reveal a mushrooming condition along with a 45° shear plane. Actual ductile failures are more commonly a combination of several of these pure mechanisms, and components of each are observable (Fig. 68).

There is a subcategory of ductile failure that is not normal overloading: creep failure. It has all of the characteristics of a ductile failure, but it occurs slowly at elevated temperatures. When a part is subjected to a static load at temperatures in the creep range for that material, it will slowly deform and break, at loads below those that would be anticipated. Creep failures can often be recognized visually by the abnormally high ductility ex-

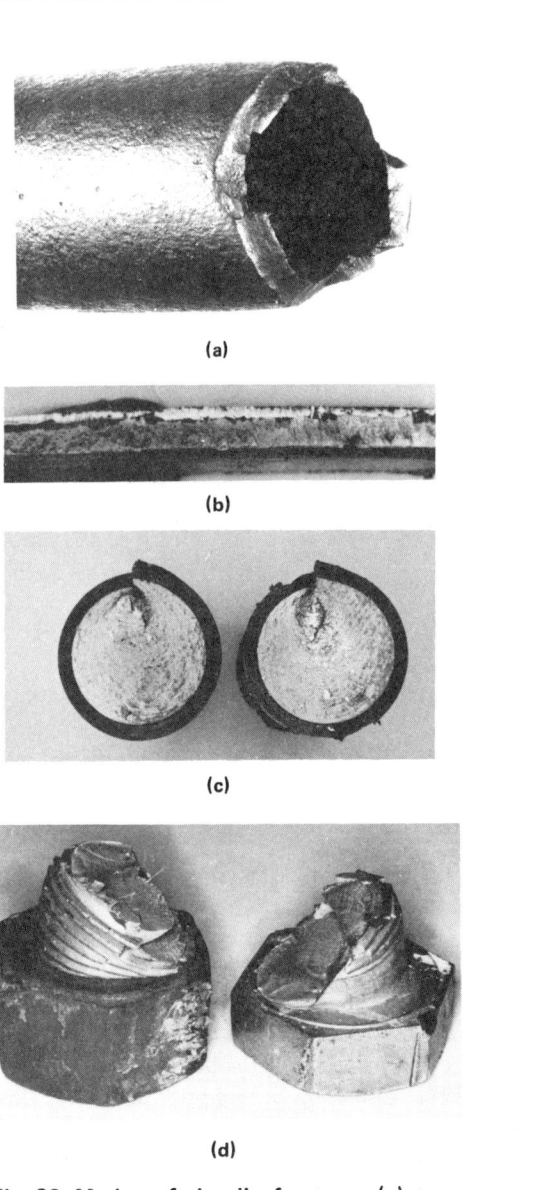

Fig. 68 Modes of ductile fracture: (a) tension; (b) shear; (c) torsion; (d) compression.

Fig. 69 Ductile creep failure of an overheated connecting-rod bolt.

hibited. Figure 69 shows a connecting-rod bolt from an overheated engine. Note that the bolt has reduced in cross section nearly to a point prior to fracturing.

Another unique characteristic observed in ductile failures is that associated with the austenitic stainless steels. Since these materials are of the face-centered cubic lattice type, they have many more slip planes than the body-centered lattice of steel. As a result, severe distortion occurs prior to failure. This is shown in the tension-test bar depicted in Fig. 70. When such characteristic distortion is viewed, the material type can be assumed to be austenitic steel.

The opposite of ductile is brittle. Brittle failures show virtually no deformation prior to failure. They are like glass shat-

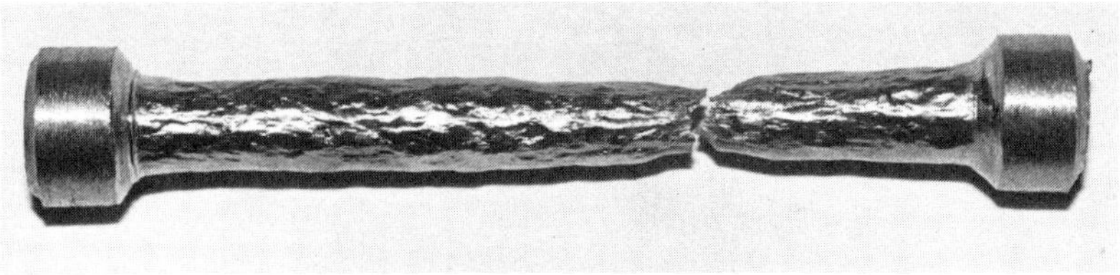

Fig. 70 Austenitic stainless steel tension-test bar showing severe slip plane distortion that occurred during tension testing.

tering. They are often catastrophic in nature, since without deformation there is no warning of impending fracture.

Brittle failures occur by either intergranular or transgranular mechanisms. Perhaps a third mechanism can be included—wherein materials have never been truly cemented together. This may be caused by excessive porosity or entrapped foreign materials. A subcategory of the transgranular fracture is fatigue, which is unique because the start-stop nature produces a characteristic fracture appearance (Fig. 71).

Electron Microscope Examination

Most fractures are readily recognizable by type, simply by visual examination at low magnification.

Sometimes, visual examination will not positively identify the fracture mode, or confirmation is needed. The electron microscope is the ultimate tool for sophisticated determinations of fracture mode. Fracture surfaces can be examined at very high magnifications, either directly or, if the part is too large, as replicas. The various modes of fracturing are readily identifiable.

Ductile fractures produce a dimpled surface. The distortion or elongating of these dimples yields information on the loading type and direction. Equiaxed dimples indicate axial tensile loading. Dimples on mating fracture surfaces elongated in the same direction indicate tensile tearing. Dimples on mating fracture surfaces elongated in opposite directions indicate shearing fracture. Torsion fractures produce a circular pattern of elongation of the dimples (Fig. 72).

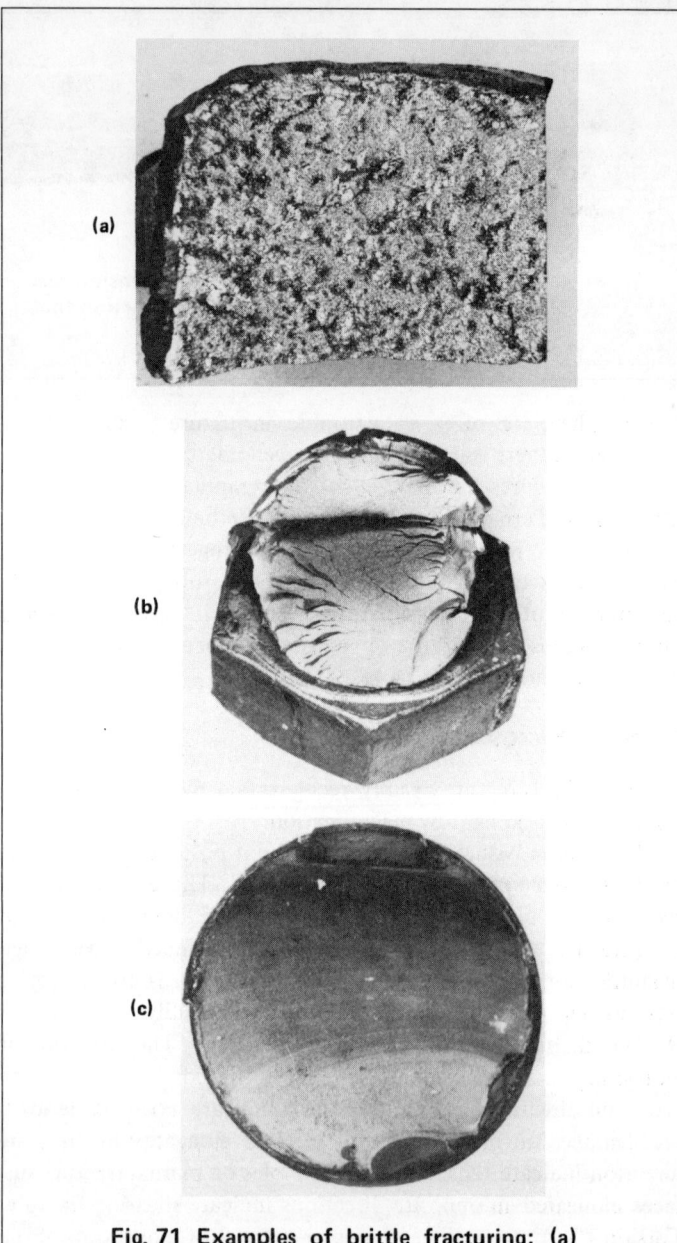

Fig. 71 Examples of brittle fracturing: (a) intergranular; (b) transgranular; (c) fatigue.

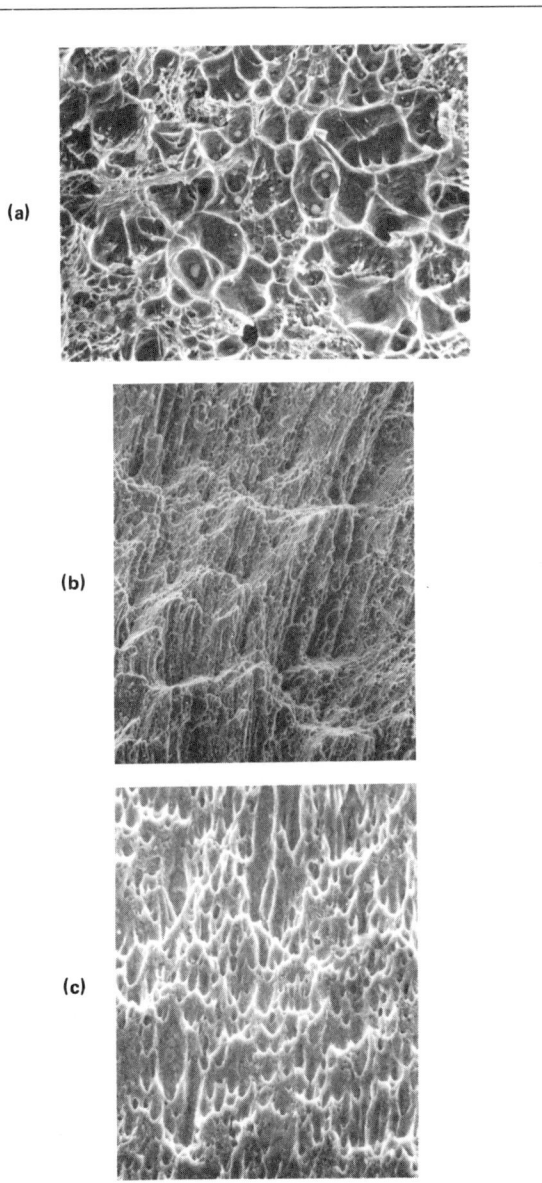

Fig. 72 Ductile fractures viewed in the electron microscope: (a) tension; (b) torsion; (c) shear. (a and c, 1000X; b, 500X)

Brittle fractures are intergranular (fracture follows the grain boundaries) or transgranular (fracture follows a cleavage pattern through the grains). Fatigue fractures are a sub-variety of the transgranular mode. Fatigue fracturing follows a start-stop path, leaving a series of parallel lines. The only other brittle-appearing fracture is caused by lack of cohesion. There are many reasons for this, but the principal ones are casting shrinkage voids, porosity, excessive nonmetallics, and lack of weld fusion. Most noncohesive fractures show evidence of other modes in conjunction. Figure 73 depicts typical brittle fracture modes.

Once the mode of fracturing is established the cause can be identified more easily. The causes of ductile failure have been discussed. The brittle failures are more difficult. Auxiliary methods may be necessary to confirm these.

Corrosion

Corrosion failures exhibit themselves in many forms. The fractures may be either intergranular or transgranular, depending on the type of corrosion. Transgranular fractures usually are indicative of higher applied stresses than the intergranular fractures. Intergranular fractures follow the more irregular path of least resistance along the grain boundaries.

Corrosion becomes evident in many different ways, some of which are not readily apparent upon visual examination. Stains, discolorations, corrosion product, and pitting are telltale signs of corrosion. Analysis of the composition of these deposits helps reveal the type. Chlorides and sulfides are common corroding elements. General pitting is characteristic of chloride attack. Moisture may result in a uniform oxidation (rusting) and wall thinning, or may produce cell-type corrosion with raised bumps with pits beneath them. High-velocity flow may produce erosion-corrosion or cavitation. Erosion-corrosion produces horseshoe-shaped pits. Cavitation produces clean, sharp-edged pits. Corrosion can be particularly severe in crevices (crevice corrosion). Dissimilar metals coupled with an electrolyte form an electrolytic cell. The least noble of the coupled metals will corrode rapidly. This is called galvanic corrosion. Stress corrosion cracking occurs in the presence of stress and corrosion, often at notches and points of high stress concentration. Caustic embrittlement is a form of stress corrosion caused by caustics. Fretting is a form of corrosion

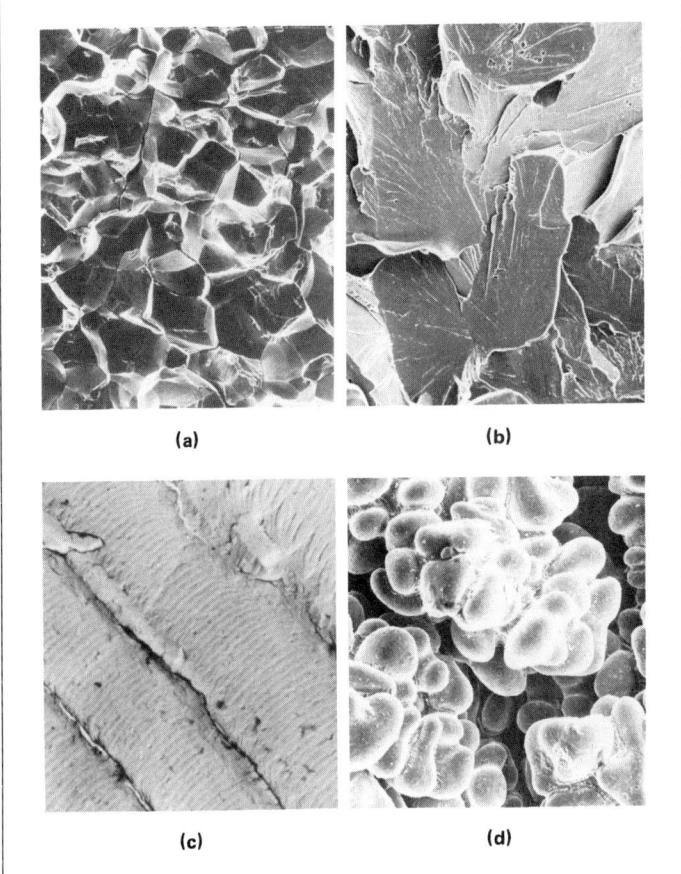

Fig. 73 Brittle fractures viewed in the electron microscope: (a) intergranular; (b) transgranular; (c) fatigue; (d) interdendritic (lack of cohesion).

produced where corrosion and metal-to-metal contact occur simultaneously. Filiform corrosion is a special form of oxygen concentration cell where long, narrow filaments of corrosion form. Dezincification corrosion occurs to brass alloys, producing localized areas with layers or plugs of friable porous copper. These are some of the principal types of corrosion. Examples are shown in Fig. 74.

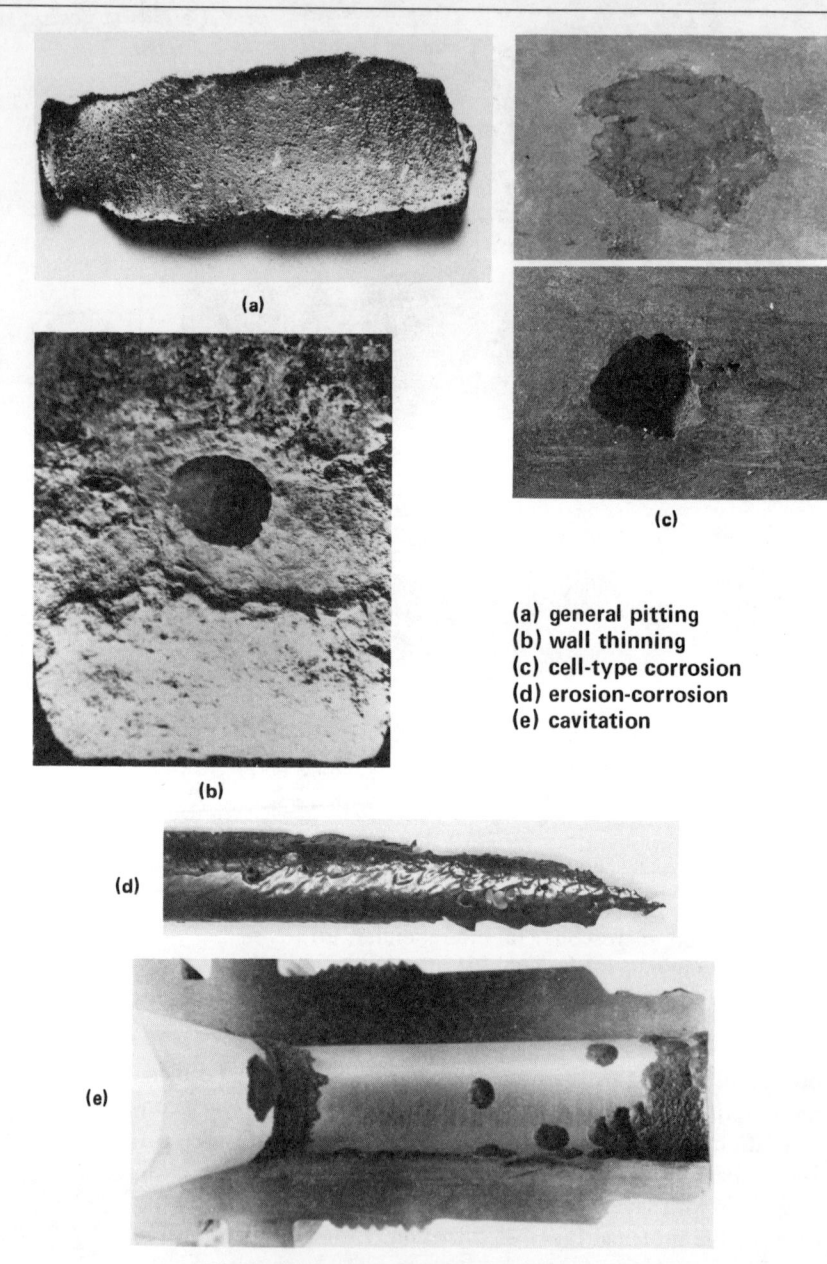

Fig. 74 Examples of corrosion failures.

(a) general pitting
(b) wall thinning
(c) cell-type corrosion
(d) erosion-corrosion
(e) cavitation

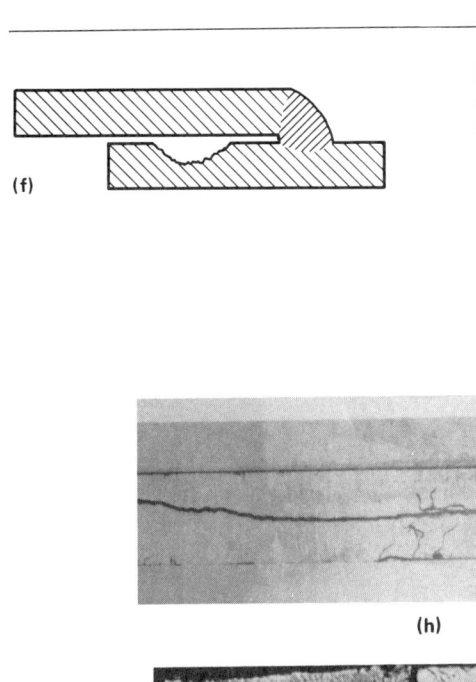
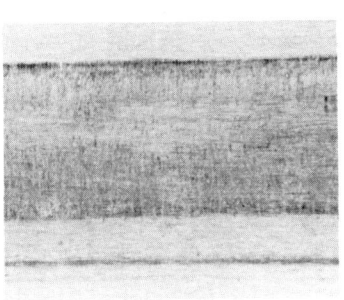
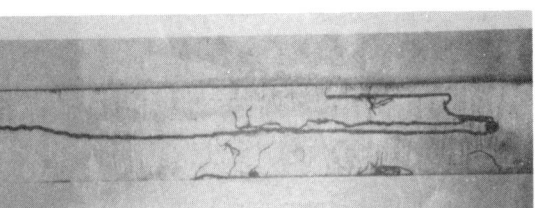

(f) crevice corrosion
(g) fretting corrosion
(h) filiform corrosion (Courtesy Oil and Gas Journal)
(i) stress corrosion cracking.

Fig. 74 continued

Fatigue

Fatigue is a very interesting type of brittle failure that reveals much information to the visual examiner. Fatigue failures occur as the result of cyclic loading and at stress levels below those required to produce failure in a single-cycle load application.

Fatigue fractures produce unique patterns of propagation. These are often described as "beach marks," "thumbnail markings," or "striations." The fracture exhibits a series of parallel lines resulting from a start-stop mechanism of crack progression (Fig. 73c).

These patterns are most often visible to the naked eye, but in some cases require the electron microscope for confirmation. They may occur as a single source of origin or with multiple origins. Careful examination of the markings can reveal information about loading characteristics (Fig. 75). Various examples are shown in Fig. 76.

Several varieties of fatigue fracturing are known to occur. Corrosion fatigue usually has multiple origins and results from corrosion pits which cause stress concentrations. Figure 77 shows a typical example.

Thermal fatigue is the result of thermal cycling. If fatigue cracks are scale-filled and blunt-ended and known to be exposed to elevated temperatures, thermal fatigue can be suspected (Fig. 78).

Other Causes of Brittle Fracture

There are many other causes of brittle failures. Heat treating may produce cracking on quenching, during heating or air cooling, or during tempering. Such cracking may exhibit heat discoloration on the fracture, quenching-oil stains, or elevated temperature scale. The characteristics may be either intergranular or transgranular, depending on material and treatment procedure (Fig. 79).

Brittle fracture is often found associated with welding. It most often occurs immediately adjacent to the weld in the heat-affected base metal (heat-affected zone). It may also occur in the weld metal itself. Visual examination can reveal much about weld failures (see Chapter 4).

Other causes of brittle failure are hydrogen embrittlement, which may be considered a form of corrosion but differs in many ways. Hydrogen embrittlement is difficult to diagnose visually. It

Failure Analysis / 131

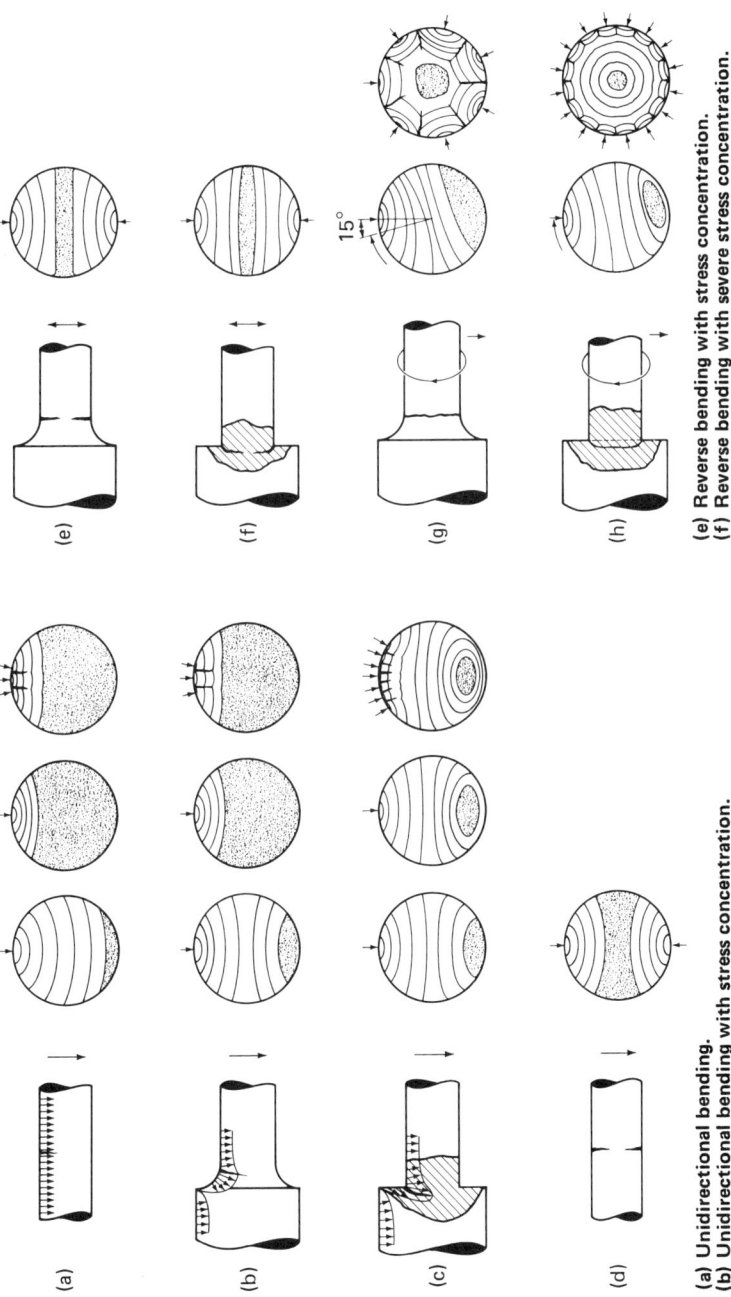

Fig. 75 Relation between fatigue fracture characteristics and type of loading. Note that multiple fractures indicate increasing stress levels. (Courtesy British Engine Insurance Ltd.)

(a) Unidirectional bending.
(b) Unidirectional bending with stress concentration.
(c) Unidirectional bending with severe stress concentration.
(d) Reverse bending.
(e) Reverse bending with stress concentration.
(f) Reverse bending with severe stress concentration.
(g) Torsional loading, smooth radius.
(h) Torsional loading, sharp corner.

132 / Inspection of Metals: Visual Examination

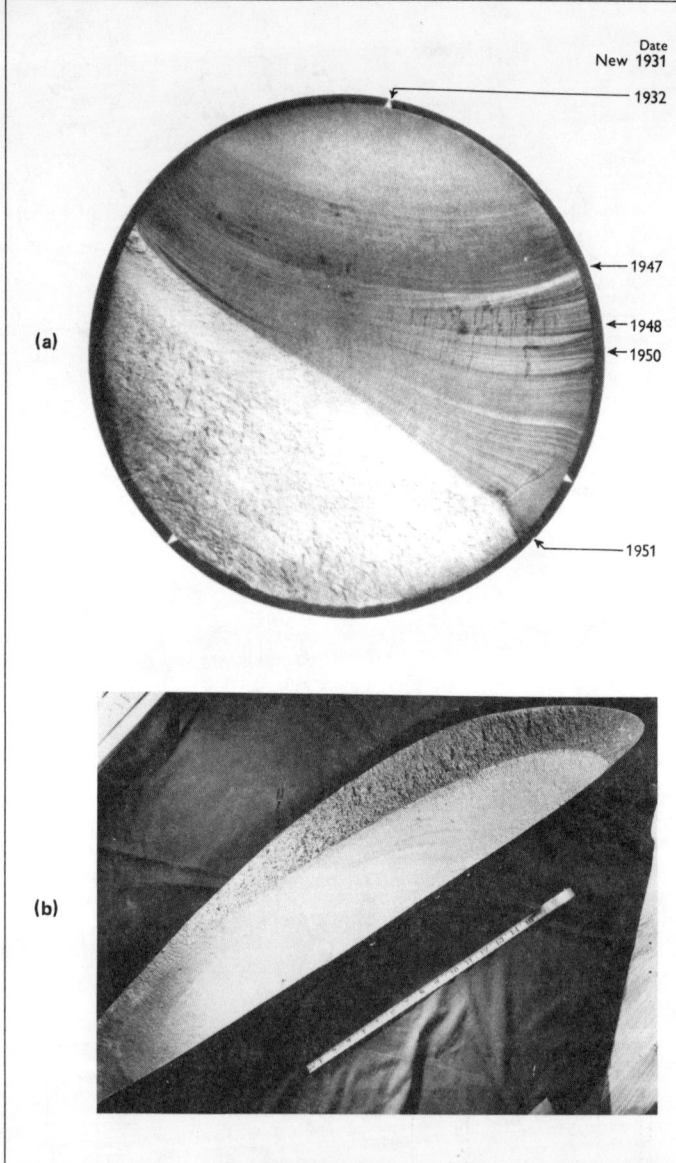

Fig. 76 Examples of fatigue failures: (a) fatigue crack that occurred over a long period of time (Courtesy British Engine Insurance

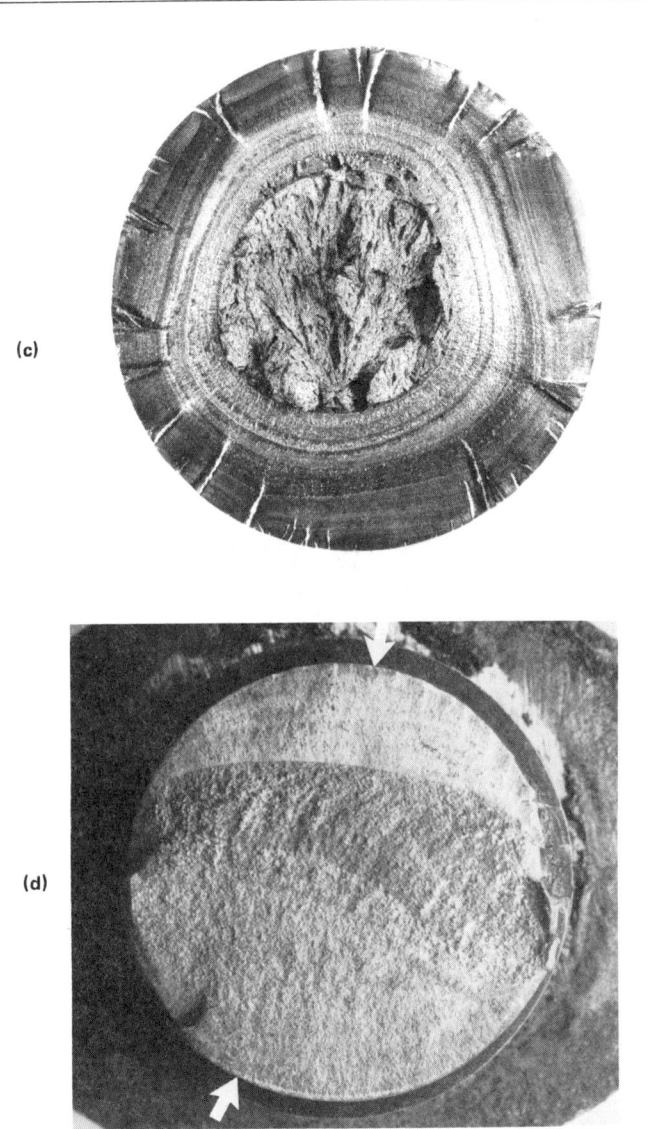

Ltd).; (b) unidirectional bending fatigue of a marine propeller; (c) torsional fatigue; (d) reverse bending fatigue.

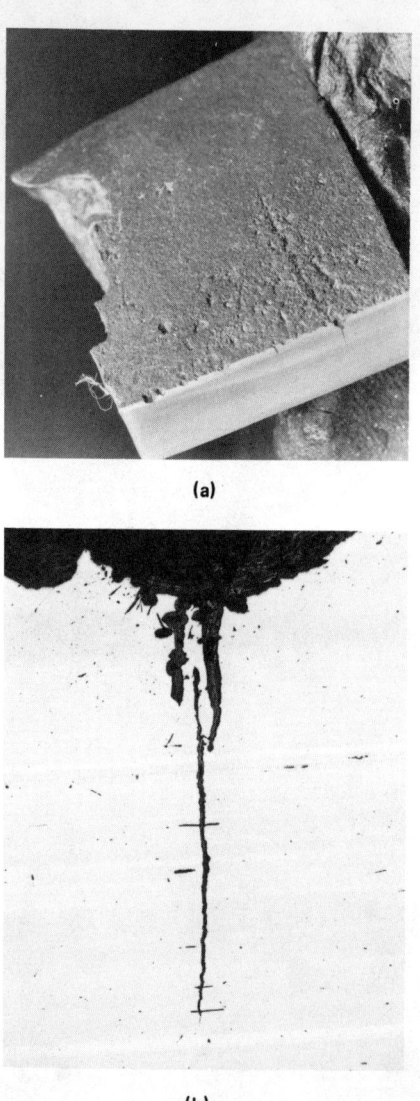

Fig. 77 Corrosion fatigue of a truck spring: (a) fractured segment of spring with one edge polished showing cracks at corrosion pits; (b) cross section through a crack at a pit.

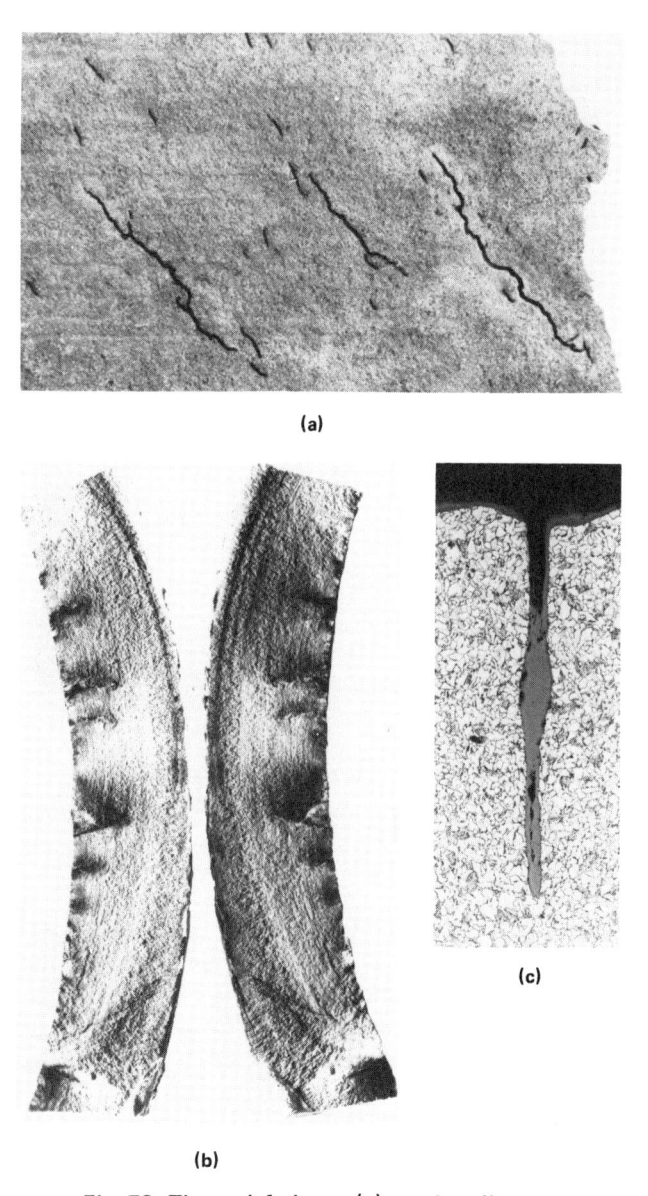

Fig. 78 Thermal fatigue: (a) cracks adjacent to fracture; (b) fracture surfaces with fatigue striations; (c) scale-filled, blunt-ended crack.

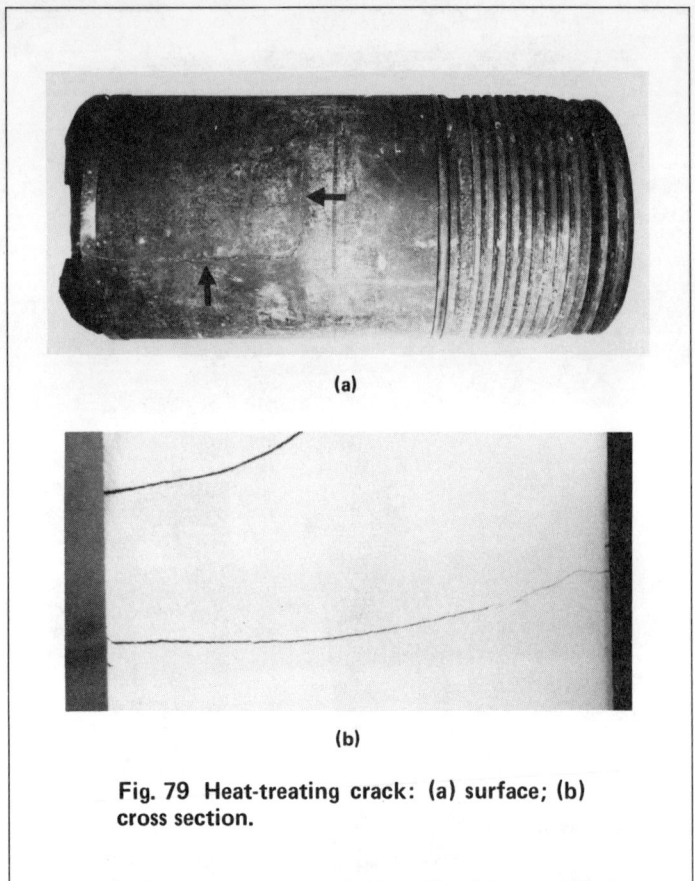

Fig. 79 Heat-treating crack: (a) surface; (b) cross section.

is sometimes accompanied by blistering (Fig. 80). It may produce obvious internal fissures observable on the fracture surface. It often initiates slightly subsurface, a characteristic best observed under the electron microscope. Another fractographic characteristic of hydrogen embrittlement is a fishhook appearance on the fracture surface (Fig. 80).

Under certain conditions, geometric shape can produce brittle-appearing failures in a ductile material. This is called triaxiality. Very sharp notches or cracks transverse to a tensile load may drastically reduce the ductility. An example is a ductile tube

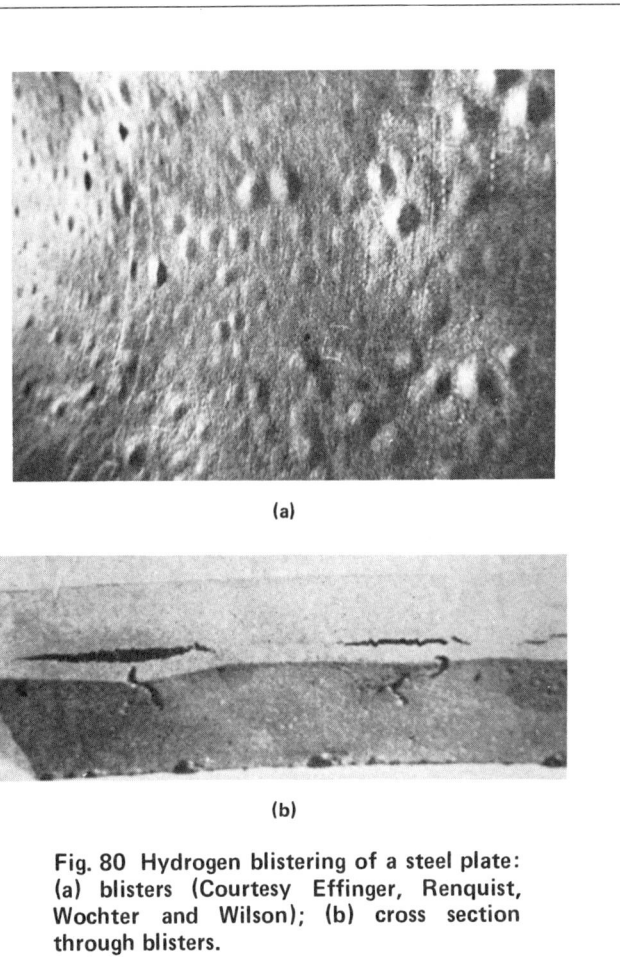

Fig. 80 Hydrogen blistering of a steel plate: (a) blisters (Courtesy Effinger, Renquist, Wochter and Wilson); (b) cross section through blisters.

pulled in tension while pressurized internally. Since the pressure opposes the normal reduction in cross section, brittle-appearing fractures result.

There are various forms of embrittlement caused by precipitation within the material and formation of undesirable phases. These are not visually apparent and require sophisticated methods to identify. They include temper brittleness, blue brittleness, sigma phase formation, and graphitization.

Impact Fracture

Materials undergo a transition from ductile to brittle behavior as the temperature is decreased. Each batch of material has its characteristic embrittling temperature. Sometimes this can occur above room temperature. Confirmation requires impact testing and cannot be determined visually. If brittle-fractured parts have been exposed to low temperatures, such exposure should be suspected as the cause of fracture.

In summary, the following are basic types and causes of fractures to look for:

Ductile
    Tension   Compression
    Torsion   Creep
    Shear

Brittle
    Manufacturing defects
    Fatigue
    Inherently brittle
    Corrosion
    Stress corrosion
    Hydrogen and caustic embrittlement
    Heat treating
    Welding
    Triaxiality
    High ductile-to-brittle transition temperature
    Precipitation and phase formation phenomena

Bibliography

Analysis of Casting Defects, 1st Edition, American Foundrymens' Association, Chicago, 1947, 133 pages.

API Bulletin on Nondestructive Testing Terminology, API Bulletin 5T1, 5th Edition, American Petroleum Institute, Dallas, Mar. 1978, 26 pages.

API Specification for Casing Tubing and Drill Pipe, Spec. 5A, 36th Edition, American Petroleum Institute, Dallas, Mar. 1982, 75 pages.

ASME Boiler and Pressure Vessel Code Section II: Material Specifications, Part A – Ferrous, 1971 Edition, American Society of Mechanical Engineers, 1971, 508 pages.

Basic Course in Optics, K. Clough, W. H. Curtin & Co., Houston, 1965.

Brazing Manual, American Welding Society, 1963, 290 pages.

A Close Look at Filiform Corrosion, J. Sudbury, O. Riggs and J. Leterle, *Oil and Gas Journal*, Sept. 8, 1958, p 118–124.

Electron Fractography Handbook, A. Phillips, V. Kerlins and B. V. Whiteson, Tech. Report ML-TDR-64-416, Research and Technology Div., Air Force Systems Command, Wright Patterson Air Force Base, Ohio, 1965, 743 pages.

Forging Handbook, W. Naujoks and D. Fabel, American Society for Metals, 1953, 630 pages.

General Catalog, Spring 1981 Edition, Metallurgical Supply Co., Houston, p 47–202.

Hydrogen Attack of Steel in Refinery Equipment, R. Effinger et al., *Proceedings of API*, API 31 M (III), 1951, p 107–33.

The Making, Shaping and Treating of Steel, H. McGannon, 9th Edition, United States Steel Corp., 1971, 1420 pages.

Mechanical and Physical Properties of Ferrous Forgings, Illinois Institute of Technology Research Institute, Committee of Hot Rolled and Cold Finished Bar Producers, American Iron and Steel Institute, New York, N.Y., 81 pages.

Metals Handbook 1955 Supplement, American Society for Metals, 1955, 207 pages.

Metals Handbook, 9th Edition, Volume 1, Properties and Selection: Irons and Steels, American Society for Metals, 1978, 793 pages.

Metals Handbook, 8th Edition, Volume 6, Welding and Brazing, American Society for Metals, 1971, 734 pages.

Metals Handbook, 8th Edition, Volume 10, Failure Analysis and Prevention, American Society for Metals, 1975, 604 pages.

Metals Handbook, 8th Edition, Volume 11, Nondestructive Inspection and Quality Control, American Society for Metals, 1976, 446 pages.

Modern Steels and Their Properties, Handbook 268, Bethlehem Steel Co., 1949, p 141.

1981 Annual Book of ASTM Standards, Part 2, Ferrous Castings; Ferroalloys, American Society for Testing and Materials, 1981, 652 pages.

1982-83 Forging Capability Chart, Forging Industry Association, Cleveland, 65 pages.

Open Die Forging Manual, 3d Edition, Forging Industry Association, Cleveland, 1982, 200 pages.

Procedure Handbook of Arc Welding Design and Practice, 10th Edition, The Lincoln Electric Co., Cleveland, 1955, 1200 pages.

Proper Identification of Bolt Heads and Nuts, N. Maycock, Steel Co. of Canada Ltd., Toronto.

Resistance Welding Manual, Volume II, 3d Edition, Resistance Welder Manufacturers' Association, Philadelphia, 1961, 266 pages.

Resistance Welding of Nickel and High Nickel Alloys, Tech. Bul. T-33, Huntington Alloy Products Div., The International Nickel Co., Huntington, W. Va., Sept. 1959, p 34.

Resistance Welding, Theory and Use, American Welding Society, 1962, 163 pages.

Schaum's Outline of Theory and Problems for Students of College Physics, 5th Edition, D. Schaum, Schaum Publishing Co., New York, N.Y., 1946, p 158-182.

Soldering Manual, American Welding Society, 1964, 170 pages.

Steel Castings Handbook, 4th Edition, C. Briggs, Steel Founders' Society of America, Rocky River, Ohio, 1970, 951 pages.

Steels for Elevated Temperature Service, 4th Edition, United States Steel Corp., 1961, p 52.

Technical Report, New Series – Volume II, British Engine Boiler and Electrical Insurance Co. Ltd., Manchester, England.

Three Keys to Satisfaction, Climax Molybdenum Co., p 11 and 12.

Tool Steel Trouble Shooter, Handbook 1964, Bethlehem Steel Corp., Bethlehem, Pa., 125 pages.

Visual Examination of Steel, G. Enos, American Society for Metals, 1940, 123 pages.

Weldability of Steels, R. Stout and W. Doty, Welding Research Council, New York, N.Y., 1953, 381 pages.

Weld Imperfections, A. Pfluger and R. Lewis, Addison-Wesley Publishing Co., Reading, Mass.

Welding Handbook, Volume 2, Welding Processes – Arc and Gas Welding and Cutting, Brazing, and Soldering, 7th Edition, American Welding Society, 1978, p 369–458

Welding Handbook, Volume 3, Welding Processes – Resistance and Solid State Welding and Other Joining Processes, 7th Edition, American Welding Society, 1980, p 1–143.

Welding Inspection, American Welding Society, 1968, 234 pages.

Welding Metallurgy, O. Henry and G. Claussen, 2nd Edition, American Welding Society, 1949, 505 pages.

Index

A

Abuse, evidence of (F), 5, 11
Arc drag, in welding (F), 89-91
Arc strikes, in welding (F), 89-91
Austenitic stainless steels, ductile failure (F), 122-123

B

Background information
 hearsay information, 110
 in failure analysis, 109-110
 operating data, 110
Bars. *See* Merchant bars
Billets. *See* Blooms, slabs and billets
Bleeder, in castings, 74
Blisters
 in castings (F), 68-69
 in plates (F), 58-59
Blooms, slabs and billets, 50
 burned steel (F), 55, 57
 cracks (F), 55, 58
 defects in (F), 55-57
 laps, 55
 rolled-in foreign material (F), 55, 57-58
 scabs (F), 55-56
 seams (F), 55-56
Blowholes, in ingots (F), 52-53
Bolt heads, identification markings (F), 5, 7-10

Borescopes (F), 45-46, 48
Brazing, 99
 defects in, 99, 100
Brittle failure
 characteristics of (F), 122-124
 compared to ductile failure (F), 119-120
 electron microscope examination (F), 126-127
 embrittlement from precipitation within, 137-138
 from heat treating (F), 130, 136
 from hydrogen embrittlement (F), 130, 137
 from triaxiality, 136-137
 from welding, 130
Buckle and kink
 in castings, 74
 in merchant bars, 61
Burn through. *See* Penetration
Burned steel
 in blooms, slabs and billets (F), 55, 57
 in merchant bars, 61
Burning, in forgings, 65

C

Camber, in merchant bars, 61
Cameras
 for field inspection kit, 111
 macro (F), 41-42

143

rapid photography, 42
    35mm (F), 40-42
    view (F), 41-42
Carbon-arc light source, 33
Castings
    defects in (F), 68-77
    entrapped gas, porosity, and blisters (F), 68-69
    hot and cold tearing and cracking (F), 69-71
    identification markings (F), 76-77
    incomplete castings (F), 73-74
    misshapen castings (F), 74, 76
    shrinkage (F), 68-70
    surface conditions, 74-75
    surface conditions caused by contamination (F), 70-73
Chromatic aberration, in lenses, 23
Clamp-offs. *See* Crushes, pushups and clamp-offs
Coddington magnifier (F), 24-25
Cold shut, in castings, 73
Cold-soldered joints. *See* Joints, cold-soldered
Contamination, in castings (F), 70-73
Correction, of lens (F), 24-25
Corrosion, 126-127
    scaling, 12, 14-15
    types of (F), 126-129
Corrosion fatigue (F), 130,134
Cracks and cracking, 15-16
    in blooms, slabs, and billets (F), 55, 58
    in castings (F), 69-71
    in flash and upset welding, 96-97, 99
    in forgings, 64-65
    in ingots, 52
    in spot welding (F), 95-96
    in welding (F), 16, 87-98
    stress concentration (F), 16, 18-19
    types of (F), 16-17
Craters and voids, in flash and upset welding (F), 96
Creep (F), 120, 122
Crushes, pushups and clamp-offs, in castings (F), 71-73
Cuts and washes, in castings, 74

D

Decarburization, in forgings, 65
Defects
    in blooms, slabs, and billets (F), 55-56
    in brazing (F), 99-100
    in castings (F), 68-77
    in electric resistance welding, 93-99
    in forgings (F), 64-68
    in ingots (F), 52-55
    in merchant bars (F), 60-61
    in plates (F), 57-60
    in soldering (F), 101-102
    in structural shapes, 60
    in tubular products, 62-63
    in welding (F), 81-93
    in wire products, 62
Denting, in tubular products, 63
Depth of field, in lenses, 24
Dial indicators (F), 37-39
Die burns, in flash and upset welding (F), 96, 98
Distortion
    in lenses, 23
    in welding, 92-93
Disturbed joints. *See* Joints, disturbed
Double plano-convex magnifier (F), 24-25
Drop, in castings, 74
Ductile failure
    causes of, 119
    characteristics of (F), 119-120
    compared to brittle failure (F), 119-120
    creep failure (F), 120-122
    electron microscope examination (F), 123, 125
    in austenitic stainless steels (F), 122-123

E

Eccentricity, in tubular products, 63
Electric resistance welding (*see also* Flash welding; Seam welding; Spot welding; Upset welding), 93
    defects in, 93-99
Electrode deposit, in spot welding (F), 93, 95-96
Electron microscope examination
    of brittle fracture (F), 126-127
    of ductile frature (F), 123, 125
Embrittlement, cause of brittle failure, 137-138
Entrapped gas. *See* Gas, entrapped
Entrapped oxides. *See* Oxides, entrapped
Equipment
    borescopes (F), 45-46, 48
    fiber optic scopes (F), 46, 48
    lighting (F), 31-34
    magnifying devices (F), 22-31
    measuring devices (F), 34-39
    mirrors (F), 45
    record-keeping mechanisms (F), 39-43

stereoscopic microscope (F), 44-45
surface finish comparators (F), 47-48
Erosion scabs, in castings (F), 72-73
Evaluation charts, for lens (F), 24, 26
Examination procedure, for failure analysis, 113-114

F

Failure analysis
   brittle failure (F), 119-120, 122-124, 130, 136-138
   corrosion (F), 126-129
   ductile failure (F), 119-125
   electron microscope examination, 123, 126
   examination procedure, 113-114
   fatigue (F), 130-133
   fracture examination on site (F), 114-115
   impact fracture, 138
   measurement, sketches, and notes, 116-117
   preparation for sample selection and removal, 112, 117-119
   record keeping, 112
   tools for, 110-111
   use of background information, 109-110
   visual fracture examination, 119-123
Fatigue (F), 127, 130
   corrosion fatigue (F), 130, 134
   relation between fatigue fracture and types of loading (F), 130-131
   thermal fatigue (F), 130, 135
   types of (F), 130, 132-133
Ferrite fingers, in forgings (F), 65-66
Fiber optics (F), 33-34
Field inspection (*see also* On-site inspection)
   camera kit for, 111
   record keeping, 112
   tool kit for, 110-111
Field metallography, 119
Finished materials. *See* Materials, finished
Fins and overfills
   in castings, 74
   in forgings (F), 65-66
   in merchant bars, 60
Fire cracks, in merchant bars, 61
Flash welding (F), 93-94
   cracking, 96-97, 99
   craters and voids, 96
   defects in (F), 96-99
   die burns (F), 96, 98
   entrapped oxides (F), 96, 98

   incomplete fusion (F), 96, 98
   problems needing mechanical adjustments, 99
Fluorescents (F), 32-33
Flux joints. *See* Joints, flux
Focal length, 22-23
Foreign materials
   entrapped in weld metal (F), 91-92
   rolled in blooms, slabs, and billets (F), 55, 57-58
Forgings, 64
   burning, decarburization, or scaling, 65
   cracking, 64-65
   defects from weld repairs, 64
   defects in (F), 64-68
   ferrite fingers (F), 65-66
   fins and overfills (F), 65-66
   flakes (F), 64-66
   identification markings, 68
   seams, 65
   shelf (F), 65-66
   shrinkage, 64
   surface laps, 65
   underfill (F), 67-68
   wrinkle (F), 67-68
Fracture, of plate (F), 116
Fracture examination, 16, 18
   on site (F), 114-115
Fusion
   in castings (F), 71
   incomplete in flash and upset welding (F), 96, 98
   incomplete in welding (F), 81, 83

G

Gages
   depth gages (F), 38-39
   hole and plug gages, 38
   radius gages (F), 38-39
   screw-pitch gages (F), 38-39
   thickness gages, 38
   thread-measuring gages (F), 38-39
Gas, entrapped, in castings, 68
Gas holes, in spot welding (F), 95-96
Gouging, in tubular products, 63

H

Halogen light source, 33
Hand-held lenses. *See* Lenses, hand-held
Hard spots, in plates, 60
Hardness, in heat-affected zones in welding (F), 91-92

Hastings triplet magnifier (F), 24-25
Hearsay information, for background information, 110
Heat-affected zones, hardness in welding (F), 91-92
Heat effects, 12
Heat treating, causes of brittle failure (F), 130, 136
Hook, in merchant bars, 61
Hydrogen blistering, brittle failure (F), 136-137
Hydrogen embrittlement, cause of brittle failure (F), 130, 137

I

Identification markings. *See* Markings
Illuminated magnifiers. *See* Magnifiers, illuminated
Impact fracture, 138
  types and causes, 138
Incandescents (F), 32-34
Inclusion, in castings, 71
Incoming materials. *See* Materials, incoming
Indentations, in spot welding (F), 93, 95
Ingots, 50
  blowholes (F), 52-53
  cracking, 52
  defects in (F), 52-55
  pipe (F), 52-54
  scabs, 52
In-process inspection, 105-106
Inspection
  at supplier's facilities, 105
  in-process, 105-106
  in the field, 106-107
  justification of, 103-104
  of finished materials, 106
  of incoming materials, 105
  on-site, 112-113
  specifications, 104-105

J

Joints, cold-soldered, in soldering (F), 101-102
Joints, disturbed, in soldering (F), 101-102
Joints, flux, in soldering (F), 101-102

K

Kink. *See* Buckle and kink

L

Laminations, in plates (F), 58
Laps
  in blooms, slabs, and billets, 55
  in forgings, 65
  in merchant bars, 61
Lenses
  chromatic aberration, 23
  Coddington magnifier (F), 24-25
  correction of (F), 24-25
  depth of field, 24
  distortion, 23
  double plano-convex magnifier (F), 24-25
  evaluation charts (F), 24, 26
  Hastings triplet magnifier (F), 24-25
  spherical aberration, 23
  types, 23
Lenses, hand-held (F), 27-28
Lighting
  carbon-arc light source, 33
  fiber optics (F), 33-34
  fluorescent (F), 32-33
  for photography, 43
  general lighting (F), 31-33
  halogen light source, 33
  incandescent (F), 32-34
  specific lighting devices (F), 33-34
Linear measuring devices (F), 34-35
Loupe, magnifying device which is an eye attachment (F), 29-31

M

Magnified examination, 3
Magnifiers, illuminated (F), 31
Magnifiers, self-supporting (F), 27, 29
Magnifiers, simple
  hand-held lenses (F), 27-28
  illuminated magnifiers, 31
  magnifying devices which are eye attachments (F), 29-31
  pocket microscopes (F), 27-28
  self-supporting magnifiers, 27, 29
Magnifying devices
  focal length, 22-23
  lens types (F), 23-26
  magnifying power, 22
  simple magnifiers (F), 27-31
Magnifying power, 22
Markings (F), 3-5
  bolt head markings (F), 5, 7-10
  date of manufacture, 3-5

identification of manufacture, 3-5
in castings (F), 76-77
in forgings, 68
in merchant bars, 62
in structural shapes, 60
material specifications, 3
number of the original heat, 3
original size of material, 3
wire rope industry markings (F), 4-6
Materials, finished, inspection, 106
Materials, incoming, inspection, 105
Measurements
for failure analysis, 116-117
use of, 18-19
Measuring devices
linear measuring devices (F), 34-35
micrometers (F), 35, 37
miscellaneous measuring devices (F), 37-39
optical comparators (F), 37-38
reticles (F), 35-36
Merchant bars, 50
buckle and kink, 61
burned steel, 61
camber, 61
defects in, 60-62
fins and overfills, 60
fire cracks and roll marks, 61
hook, 61
identification markings, 62
laps, 61
pipe, 61
rolled-in scale, 61
scratches, 61
seams, 61
shear distortion, 61
slivers, 60
twist, 62
underfills, 60
Metallography. See Field metallography
Micrometers (F), 35, 37
inside micrometers (F), 38-39
tubing wall measuring (F), 38-39
Microscopes, pocket (F), 27-28
Mirrors (F), 45
Misrun, in castings (F), 73
Misshapen weld, in spot welding (F), 93, 95
Motion pictures. See Photography

N

Naked eye examination, 2
Notes, for failure analysis, 116

O

On-site inspection, 112-113
Operating data, for background information, 110
Optical comparators (F), 37-38
Overfills. See Fins and overfills
Oxides, entrapped in flash and upset welding (F), 96, 98

P

Penetration
excess in welding (F), 91
insufficient in welding (F), 83-84
Photography
for record-keeping (F), 40-42
lighting for, 43
macro cameras (F), 41-42
motion pictures, 43
rapid photography cameras, 42
35mm cameras (F), 40-42
view cameras (F), 41-42
Pipe
in ingots (F), 52-54
in merchant bars, 61
Plate, in castings, 74
Plates, 51
blisters (F), 58-59
defects in (F), 58-60
hard spots, 60
inspection of, 57-58
laminations (F), 58
scale, 58
Pocket microscopes. See Microscopes, pocket
Porosity
in castings (F), 68-69
in welding (F), 86-87
Protractors, 38
bevel protractors, 38
Pull downs, in castings, 74
Purpose of visual examination, 1-2
Pushups. See Crushes, pushups and clamp-offs

R

Rat, in castings, 71
Rattails, in castings, 74
Record keeping
for field inspection, 112
in visual examination, 20

mechanisms for, 39-43
motion picture photography, 43
photography (F), 40-42
photography lighting, 43
tape recorders (F), 40-41
video recorders, 43
Replication techniques, 118-119
Results, of visual examination, 20
Reticles (F), 35-36
Roll marks, in merchant bars, 61
Rosin joints. *See* Joints, flux
Runouts, in castings (F), 74

S

Sag, in castings, 74
Samples
  preparation for removal, 112, 117-119
  selection of, 117-119
Scabs (*see also* Erosion scabs)
  in blooms, slabs, and billets (F), 55-56
  in castings (F), 74-75
  in ingots, 52
Scales and scaling
  corrosion, 12, 14-15
  in forgings, 65
  in plates, 58
  rolled-in in merchant bars, 61
  temperatures at which it becomes appreciable for steel (F), 12-14
Scar, in castings, 73
Scratches, in merchant bars, 61
Seam welding (F), 93-94
Seams
  in blooms, slabs, and billets (F), 55-56
  in forgings, 64
  in merchant bars, 61
Self-supporting magnifiers. *See* Magnifiers, self-supporting
Shear distortion, in merchant bars, 61
Shelf, in forgings (F), 65-66
Shifts, in castings (F), 74, 76
Short pour. *See* Misrun
Shrinkage
  in castings (F), 68-70
  in forgings, 64
Simple magnifiers. *See* Magnifiers, simple
Sketches and sketching, 3
  for failure analysis, 116-117
Slabs. *See* Blooms, slabs and billets
Slag
  in castings (F), 71-72
  in welding (F), 85-86

Slivers, in merchant bars, 60
Solder, excess and insufficient (F), 101-102
Soldering
  cold-soldered joints (F), 101
  defects in (F), 101-102
  disturbed joints (F), 101
  excess and insufficient solder (F), 101-102
  flux joints (F), 101-102
  list of materials that can be soldered, 99-100
Specifications, 104-105
Spherical aberration, in lenses, 23
Spot welding (F), 93-94
  cracks and gas holes (F), 95-96
  defects in (F), 93, 95-96
  indentations (F), 93, 95
  misshapen weld (F), 93, 95
  surface fusing and/or electrode deposit (F), 93, 95-96
Steel color code chart, *inside insert facing page 54*
Steels, temperature at which scaling becomes appreciable (F), 12-14
Stereoscopic microscopes (F), 44-45
Sticker. *See* Rat
Stress concentration, cracking (F), 16, 18-19
Structural shapes, 51
  defects in, 60
  identification markings, 60
Supplier's facilities, inspection, 105
Surface finish comparators (F), 47-48
Surface fusing, in spot welding (F), 93, 95-96
Swell, in castings, 74

T

Tape recorders, for record-keeping (F), 40-41
Tearing, in castings (F), 69-71
Temper colors, *insert facing page 54*
Thermal fatigue (F), 130, 134
Thought transference, 3
Triaxiality, cause of brittle failure, 136-137
Tubular products, 51
  defects in, 62-63
  denting and gouging, 63
  eccentricity, 63
  mill-type defects, 63
  weld defects, 63
Twist, in merchant bars, 62

U

Undercutting, in welding (F), 83-85
Underfills
 in forgings (F), 67-68
 in merchant bars, 60
Upset welding (F), 93-94
 cracking, 96-97, 99
 craters and voids, 96
 defects in (F), 96-99
 die burns (F), 96, 98
 entrapped oxides (F), 96, 98
 incomplete fusion (F), 96, 98
 problems needing mechanical adjustments, 99
 surface indications of weld quality (F), 97

V

Veins, in castings (F), 74-75
Video recorders, 43
Visor types, magnifying devices which are an eye attachment (F), 29-30
Visual examination
 evidence of abuse (F), 5, 11
 identification markings (F), 3-10
 magnified examination, 3
 of corrosion scaling, 12, 14-15
 of cracking, 15-16
 of heat effects, 12
 prior to welding, 80-81
 purpose, 1-2
 results and record keeping, 20
 sketching, 3
 thought transference, 3
 use of measurements, 18-19
 use of naked eye, 2

Voids. *See* Craters and voids

W

Warpage, in castings, 74
Washes. *See* Cuts and washes
Weld defects
 in forgings, 65
 in tubular products, 63
Weld splatter, in welding (F), 89-91
Welding (*see also* Electric resistance welding), 79
 arc strikes, arc drag, and weld splatter (F), 89-91
 cause of brittle failure, 130
 cracking (F), 16, 87-89
 defects in (F), 81-93
 distortion, 92-93
 entrapped foreign material in weld metal (F), 91-92
 excess penetration (F), 91
 hard heat-affected zones, 91-92
 incomplete fusion (F), 81, 83
 insufficient penetration (F), 83-84
 porosity (F), 86-87
 slag (F), 85-86
 undercutting (F), 83-85
 visual examination prior to (F), 80-81
Wire and wire products, 51
 defects in, 62
Wire rope industry, identification markings (F), 4-6
Wrinkle
 in castings (F), 74-75
 in forgings (F), 67-68
Wrought materials
 defects in, 52-63
 types of, 50-52